Competences in context

Knowledge and Capacity Development in Public Water Management in Indonesia and the Netherlands

Competences in Context

Knowledge and Capacity Development in Public Water Management in Indonesia and the Netherlands

DISSERTATION

Submitted in fulfilment of the requirements of
the Board for Doctorates of Delft University of Technology
and of
the Academic Board of the UNESCO-IHE Institute for Water Education
for the Degree of DOCTOR
to be defended in public
on Tuesday, April 23, 2013, at 15.00 hours
in Delft, The Netherlands

by

Judith Machteld Kaspersma

born in Winschoten, the Netherlands

Master of Science in Irrigation and Water Engineering
Wageningen University, the Netherlands

This dissertation has been approved by the supervisor
Prof. dr. ir. G.J.F.R. Alaerts

Dr. Ir. J.H. Slinger, copromoter

Members of the Awarding Committee:

Chairman	Rector Magnificus TU Delft, the Netherlands
Prof. Dr. Ir. Szöllösi-Nagy	Chairman, UNESCO-IHE
Prof. dr. ir. G.J.F.R. Alaerts	Promoter, UNESCO-IHE/TU DELFT
Dr. Ir. J.H. Slinger	Copromoter, TU DELFT
Prof. dr. ir. W.A.H. Thissen	TU DELFT, The Netherlands
Prof. dr. ir.H.H.G. Savenije	TU DELFT, The Netherlands
Prof. dr. ir. C.A.M. Termeer	Wageningen University, The Netherlands
Dr. Ir. R. Sjarief	Bandung Institute of Technology, Indonesia
Prof. dr. ir. P. van der Zaag	UNESCO-IHE/TU DELFT, reserve member

CRC Press/Balkema is an imprint of the Taylor & Francis Group, an Informa business

Picture on the front page by Melchert Meijer zu Slochtern

Published by:
CRC Press/Balkema
PO Box 11320, 2301 EH Leiden, The Netherlands
e-mail: PUB.NL@taylorandfrancis.com
www.crcpress.com – www.taylorandfrancis.co.uk

ISBN: 978-1-138-00097-1

Summary

International cooperation for reaching development goals has expanded gradually since the 1950s. The effectiveness of the Overseas Development Aid (ODA) has become a topic of great public interest and the sixty years of work can be viewed as a long-term societal learning process about what development really is (Zevenbergen and Boer, 2002). Much of the public debate amongst non-economists takes for granted that, if the funds were made available, poverty would be eliminated, and at least some economists, notably Sachs (2005) agree (Deaton, 2010). Others, most notably Easterly (2009) believe that a bottom-up approach, not necessarily involving large funds, but giving a voice to local communities to indicate their needs themselves would be much more successful. A growing body of experience exists to demonstrate that finance alone cannot do it, and capacity and knowledge are increasingly seen as the constraints to appropriate decision making, absorption of funds, and effective results on the ground. According to the Paris Declaration on Aid Effectiveness (OECD, 2005), development efforts in many of the poorest countries will fail, even if they are supported with substantially increased funding, if the development of sustainable capacity is not given greater and more careful attention.

Because of its complexity, and because of its common-property and distributed nature with many stakeholders, the water sector is particularly dependent on effective institutions and, therefore, on strong capacities and a solid knowledge base at the individual and institutional levels (Cosgrove and Rijsberman, 2000). It is therefore not surprising that the water sector was one of the first sectors to introduce knowledge and capacity development (KCD) initiatives (Alaerts et al., 1991; Hamdy et al., 1998; Appelgren and Klohn, 1999; Downs, 2001; Bogardi and Hartvelt, 2002; FAO, 2004; Alaerts, 2009b; Whyte, 2004). However, despite the attention for KCD, managing water systems and providing water services to citizens remain daunting challenges.

The aim of this thesis is to deepen understanding of the dynamic process of knowledge and capacity development and the numerous contextual factors that influence capacity from the individual to the system level, so as to improve the effectiveness of KCD programs and interventions.

The reviewed literature reveals that a tension exists between models of KCD that are usable and work in practice, and models that are more complex and reflect reality more fully, but are difficult to apply in practice. The adapted KCD conceptual model that I adopt (Alaerts and Kaspersma, 2009) is comprehensive in the sense that it views KCD at the individual, organisational and institutional levels simultaneously and acknowledges the interaction between these levels. Other models take into account the influence of levels other than the primary level of analysis, but do not assess the existing knowledge and capacity at those levels as well. They focus on the capacity at one level only. The adapted KCD conceptual model serves as a basis, and ordering framework, in the investigation of KCD in the water sector. In addition I draw upon theory from the fields of human resource development, learning, organisation and management sciences and policy analysis to explain the relations between different components of the KCD system (Chapter 2). At the individual level I adopt theory on professional competence (Cheetham and Chivers, 2005; Sultana, 2009; Oskam, 2009) to explain the composition of knowledge and capacity at the individual level and the combination of different competences required by water professionals (Chapter 6). At the level of the organisation, I use Burns and Stalker's classification of mechanistic and organic organisational structures (1961) and Mintzberg's structure in fives theory (1980) to explain how formal organisational structure influences KCD (Chapter 5). At the level of the institutions I draw upon theory on advocacy coalitions (Sabatier and Jenkins-Smith, 1993) and the multiple streams framework developed by Kingdon (1995) to explain how coalitions continuously need to promote their agendas, which embody new knowledge and capacities, in order to influence existing policy regimes supported by the establishment. A window of opportunity (Kingdon, 1995), often brought about by external events that trigger a political reaction, is required to create a transition point to a new paradigm, which allows the inclusion of new knowledge and capacities (Chapter 4).

I apply the adapted framework and additional supportive theory to study KCD in two public sector organisations, considered representative of the water sector, over a longer period of time, within their institutional contexts. The first case is the Directorate General of Water Resources (DGWR) of the Ministry of Public Works (MPW) of Indonesia. In this case I choose to assess KCD by studying local and international post-graduate education (IPE). I hypothesise that IPE is relatively important in a society where few other KCD mechanisms are assumed to be available. In many developing countries and countries in transition, IPE is an important means for accessing knowledge that is not available locally. The second case is the executive arm of the Dutch Ministry of Infrastructure and Environment, the Rijkswaterstaat. This case was chosen, because I wanted to investigate how knowledge and capacity develop and influence decision making in an organisational unit similar to the DGWR, but located in a relatively well developed economy where I hypothesise that a wide array of KCD mechanisms is available to generate and exchange new knowledge.

A mixed method approach is adopted using surveys and semi-structured interviews in both cases to analyse how water professionals acquire knowledge and develop capacities, and I undertake a historical analysis of both the Indonesian water sector and the Dutch water sector to study how the cultural and environmental features and priorities in society at distinct junctures in time have influenced the use of certain KCD mechanisms. In the historical analysis, I identify three distinct phases in each case, as illustrated in Table 1.1, which are characterised by coherent paradigms. The phases cover about 40 years. The introduction of these phases is necessary to allow analysis of the same system under different circumstances, and with different institutional and contextual parameters. The methodological differentiation of respondents in the Indonesian case as function of their local post-graduate education (LPE) or international post-graduate education (IPE) experience, and for both cases the differentiation of the administrative and political context in the country and sector per phase, proved useful in generating more detailed insights in the development of the competences in the Indonesian and Dutch water sector, over half a century, within the evolving economic, administrative and political contexts. In both cases I search for systemic parallels and differences in order to infer potentially general rules for KCD processes.

First, at the institutional level, the differences between the cases lie in the political leverage and in the way the advocacy coalitions managed to argue their agendas. In the Indonesian case knowledge brought in, among others, by the donor community, international consultants and reform-minded officials, was perceived as threatening the status quo in the DGWR, just like the Rijkswaterstaat felt challenged at some point in the 1970s by the new concepts embodied in the positions of the environmental lobby. In the Indonesian case however, the coalition did not manage to shift the traditional technocratic water policies to a more integrated management of water resources (IWRM) during Phases I and II (roughly from 1970 to 1998), because there was no leverage for IWRM and IWRM would require DGWR to delegate powers. The transition from Phase II to III provided the window of opportunity to shift to a new paradigm, featuring more decentralised decision-making, more accountability and introduction of IWRM principles in sectoral policies and practice. Similarly, I observe in the Dutch case, in the same transition period, the increased attention for accountability and transparency, and the struggle to manage the outsourcing process that was introduced partly as a policy to slim down government and partly to compensate the lack of technical competence in the organisation.

Table S.1. General description of transitions and institutional paradigmatic phases in the Indonesian and Dutch cases

Indonesia case	The Netherlands case
Phase I (1970 - 1987) Development of infrastructure Technocratic Little space for other professions than engineering in water management Organisation based on seniority	**Phase I (1950 – 1970)** Development of infrastructure Technocratic Little space for other professions than engineering in water management
Phase II (1987 - 1998) Increasingly authoritarian state – loyalty to regime increasingly important Failing effort to implement IWRM	**Phase II (1970 - 2002)** Environmental values incorporated by left-wing government Increasing role of civil society and stakeholder participation IWRM becomes policy
Phase III (1998 -) Reformation leading to decentralisation Increased transparency and accountability Law 7 on IWRM Increasing interest in governance competence	**Phase III (2002 -)** Privatisation for increased transparency and efficiency of government. Organisation becomes more hierarchic

Secondly, at the organisational level, the organisational structure of the Rijkswaterstaat was due to become more organic (in the sense of Burns and Stalker) as it evolved from Phase I to Phase II to become able to deal with a wider array of disciplines, professions and knowledge within its walls. This was a direct result of the changes in the institutional environment. In the Indonesian case the organisational structure was found to remain largely unchanged because at the institutional level the leverage was limited for change to an IWRM and governance oriented paradigm. In Phase III (from 1998 onwards), the organisation continues to be marked by a high degree of formalisation and centralisation of responsibilities, expressed in a strong hierarchy and division of labour and routines, even though the water management challenges increasingly demand interdisciplinary knowledge and capacities that are to be found outside the organisation. A more organic structure would facilitate the exchange of knowledge with actors possessing other knowledge, but also requires an acknowledgement that such knowledge cannot be found solely at the top or even within the hierarchy. Interestingly, in the Dutch case in Phase III, the global tendency to seek more transparency and (budget) accountability led to an increasingly mechanistic model of organisational structure, with more central control over budgets, work plans and Human Resource Management (HRM), rather than more decentralised decision-making. This shift may have worked well for accountability and transparency, but it did create an atmosphere in the organisation that was less conducive to creating and exchanging knowledge among staff; staff grew reluctant to take initiative.

Third, at the individual level, it became clear that in Phase I and for a large part in Phase II as well, both cases showed a firmly 'technical' default orientation, that in the Indonesian case also transpired in the choices for training and post-graduate education. During Phase II and Phase III, in both cases the line civil servants have developed strong preferences for administrative skills causing the substantive technical knowledge to become weaker, or at least less prominent, in the skills mix. Governance competence in the Dutch case in Phase II increased, as the water professionals needed to acquire the knowledge and skills to work with multiple stakeholders. In the Indonesian case during the same Phase and time period this could not yet happen as the political and institutional regime became more authoritarian, rewarding loyalty at the cost of competence.

Both cases underline the importance yet the low valuation placed on tacit knowledge. In both cases tacit knowledge is lost through outsourcing and lack of succession planning, whereas individual respondents and interviewees systematically highlight the importance of tacit knowledge to do their work. The Indonesian case study suggests strongly that tacit knowledge is the most important result gained from IPE, and, by extension, other international long-term exposure. This study reveals that tacit knowledge needs to be addressed explicitly in organisations, in formal KCD mechanisms such as education and training, by arranging succession planning, by providing the opportunity to enter mentor-coach relations, but also by creating more informal opportunities, incentives and an atmosphere for knowledge exchange.

Finally, the study confirmed that a conceptual model of KCD should accommodate the existence of three nested levels, namely the individual, organisational and institutional levels, and be able to clarify the inter-relations between these three different levels. In this thesis I revised and expanded an existing model to obtain an adapted KCD conceptual model and subsequently complemented this by an analysis of the dynamic nature of the interactions between the different levels, facilitated by supportive theories. The adapted KCD conceptual model can be applied together with the supporting theories as an ordering and explanatory framework. Further research is suggested in other sectors and case studies for validation and completing the KCD conceptual model and its new arrangement. Research could, for example, focus on the effect of personality and attitude on individual capacity development, and leadership as a personal competence and as a specific component of organisational capacity.

CONTENTS

TABLE OF FIGURES

LIST OF TABLES

LIST OF ABBREVIATIONS

5Cs	5 Core Capabilities
ACF	Advocacy Coalition Framework
ADB	Asian Development Bank
ANOVA	Analysis of Variance
AWT	Advisory Committee for Science and Technology Policy
Balai PSDA	Basin Water Resources Management Units
BAPPENAS	Central Planning Agency of the Government of Indonesia
BBWS	Strategic River Basin Management Units
BWS	River Basin Management Units
CDRF	Capacity Development Results Framework
CFO	Chief Financial Officer
CLC	Corporate Learning Centre
CMC	Corporate Mobility Centre
CPB	Capacity Building Project
CSO	Civil Society Organisation
DG	Director-General
DGWR	Directorate General of Water Resources of the Ministry of Public Works in Indonesia
DISC model	Dominance – Influence – Steadiness – Compliance quadrant behavioural model
ECDPM	European Centre for Development Policy Management
ERR	Economic Rate of Return
FAO	Food and Agricultural Organisation of the United Nations
GDP	Gross Domestic Product
GLOBE	Global Leadership and Organisational Behaviour Effectiveness
GTI	Large Technological Institute
HCA	Human Capabilities Approach
HID	Chief Engineer-Director/Managing Director
HRD	Human Resource Development
HRM	Human Resource Management
IAD	Institutional Analysis and Development
ICS	Information and Communication Systems
IMF	International Monetary Fund
IMT	Irrigation Management Transfer
IOB	Policy and Operations Evaluation Department of the Ministry of Foreign Affairs of the Netherlands
IOMP	Irrigation Operation and Maintenance Policy
IPE	International Post-graduate Education
IWA	International Water Association
IWRM	Integrated Water Resources Management
JIWMP	Java Integrated Water Management Project
KCD	Knowledge and Capacity Development
KIVI-NIRIA	Royal Institute of Engineers

KNW	Royal Dutch Water Network (new name after the merger of NVA and KVWN)
LenCD	Learning Network on Capacity Development
LP3ES	Institute for Social and Economic Research, Education and Information
LPE	Local Post-graduate Education
MDG	Millennium Development Goal
M&E	Monitoring and Evaluation
MPW	Ministry of Public Works in Indonesia
MSF	Multiple Streams Framework
NGO	Non-Governmental Organisation
NHV	Dutch Hydrological Society
NRLO	Dutch Council for Agricultural Research
NVA/KVWN	Royal Dutch Water Network
NWP	Netherlands Water Partnership
ODA	Overseas Development Aid
OECD	Organisation for Economic Cooperation and Development
O&M	Operation & Maintenance
RBM	Results Based Management
RBT	River Basin Territory
RIKZ	National Institute for Coast and Sea
RIZA	National Institute for Freshwater Management
RMNO	Council for Environmental and Nature Research
SPSS	Statistical Data Package for the Social Sciences
TKPSDA	River Basin Council
TNO	Netherlands Organisation for Applied Scientific Research
TNO-NITG	TNO-Central Geoscientific Information and Research Institute
UK	United Kingdom
UNDP	United Nations Development Programme
UNESCO	United Nations Educational, Scientific and Cultural Organisation
Unesco-IHE	UNESCO - IHE Institute of Water Education
Unesco-IHP	UNESCO - International Hydrological Programme
USA	United States of America
WATSAL	Water Resources Sector Adjustment Loan
WB	World Bank
World Bank – IEG	World Bank Independent Evaluation Group
WRM	Water Resources Management
WUA	Water Users Association

1 Introduction

1.1   INTERNATIONAL DEVELOPMENT

According to the United Nations Development Programme (UNDP), human development is about creating an environment in which people can develop their full potential and lead productive, creative lives in accord with their needs and interests. Human development is thus about expanding the choices people have to lead lives that they value. Amartya Sen (1999) similarly mentions 'expanding the freedoms that people enjoy'. Development thus is about more than economic growth, which is only a means — albeit a very important one —of enlarging people's choices (UNDP, 2008b).

Fundamental to enlarging the choices of people is the development of human capabilities—the range of things that people are able to do or be in life. The most basic capabilities for human development are to lead long and healthy lives, to be sufficiently knowledgeable in order to have access to the resources needed for a decent standard of living and to be able to participate in the life of the community. Without these, many choices are simply not available, and many opportunities in life remain inaccessible (UNDP, 2008b).

Since the 1950s international cooperation in reaching development goals has gradually grown. The effectiveness of the Overseas Development Aid (ODA) has become a topic of great public interest and the sixty years of work have most of all been a long-term learning process about what development really is (Zevenbergen and Boer, 2002).

Much of the public debate among non-economists takes for granted that, if the funds were made available, poverty would be eliminated, and at least some economists, notably Sachs (2005) agree (Deaton, 2010). Others, most notably Easterly (2009) are deeply sceptical. Easterly believes that a bottom-up approach, not necessarily involving large funds, but giving a voice to local communities to indicate their needs themselves would be much more successful. The debate runs the gamut from the macro level – can foreign assistance raise growth rates and eliminate poverty? – to the micro level – what sorts of projects are likely to be effective? Should aid focus on electricity and roads, or on the provision of schools and clinics? (Deaton, 2010). It is clear from literature that we don't have a full understanding yet of what actually works and what does not (Deaton, 2010). Much of the shortfall is attributed by donors and by countries to inadequate development of local knowledge and capacities. A growing body of experience exists to demonstrate that finance alone cannot do it; capacity and knowledge are increasingly seen as the constraints to proper decision making, adsorption of funds, and effective results on the ground. The OECD stresses that adequate country capacity is one of the critical missing factors in current efforts to meet the Millennium Development Goals (MDGs) (2006). Development efforts in many of the poorest countries will fail, even if they are supported with substantially increased funding, if the development of sustainable capacity is not given greater and more careful attention. This has also been articulated in the 2005 "Paris and Accra Declarations on Aid Effectiveness (OECD, 2008). A growing body of literature on Knowledge and Capacity Development (KCD) has started to emerge since the beginning of the 1990s (UNDP, 1997; OECD, 2006; United Nations, 2006; World Bank - IEG, 2008) and almost a quarter of ODA is devoted to KCD now (Whyte, 2004). In this chapter I will first discuss how capacity and its development are defined, then briefly introduce the opinions about KCD measurement.

1.2   KNOWLEDGE AND CAPACITY DEVELOPMENT

Different authors define capacity according to the theoretical (or political) frameworks from which they approach the subject. This means that there is no completely common capacity language or set of terms to help communication and shared learning on capacity (Ubels, 2010; Baser and Morgan, 2008; Morgan, 2006; Brown et al., 2001; Brinkerhoff, 2005). In 1993 Morgan defined capacity as 'the ability of individuals, groups, institutions and organisations to identify and solve development problems over time' (Morgan, 1993). In contrast, Hildebrand and Grindle (1994) focus on the dynamic aspects of capacity, stating 'Capacity is the ability to perform appropriate tasks effectively, efficiently, and sustainably'. This implies that capacity is not a passive state - the extent of Human Resource Development (HRD), for example - but part of an active process'. Alaerts (1999) later amalgamated these concepts in his definition of capacity for the water sector, stating: 'Capacity allows individuals, organisations and relevant institutions to perform in such a way that the sector as an aggregate can perform optimally, now as well as in the future'. All definitions make clear, however, that KCD as a concept is applicable to developing countries, but pertains equally to the more developed and richer economies.

1.3   WATER[1] MANAGEMENT

Because of its complexity, the water sector is particularly dependent on effective institutions and, therefore, on strong capacities and a solid knowledge base at the individual and institutional levels (Cosgrove and Rijsberman, 2000). It is therefore not surprising that the water sector was one of the first sectors to introduce KCD initiatives (Alaerts et al., 1991; Hamdy et al., 1998; Appelgren and Klohn, 1999; Downs, 2001; Bogardi and Hartvelt, 2002; FAO, 2004; Alaerts, 2009b; Whyte, 2004).

In the past, water resources management focused on well-defined problem; problems that had grown increasingly urgent during the 19th and 20th century, such as urbanisation, and intensified industrial and agricultural productivity. The sea and rivers had to be controlled to protect countries from flooding. Many of these issues were dealt with in isolation, and potentially undesirable long-term consequences were not taken into consideration. This has been characterised as a vertical, command and control approach (Pahl-Wostl et al., 2007). The past three decades, on the other hand, have seen the development of a more integrated approach to the water system, with more attention for physical and institutional issues inextricably bound up with water.

Water is not solely a local, national or regional issue that can be governed at any of these levels alone. On the contrary, global interdependencies weave through water issues, and decisions relating to water use on a local, national, river-basin or regional level often cannot be isolated from global drivers, trends and uncertainties (World Water Assessment Programme, 2012). Similarly, water issues are not confined to one sector; instead their governance requires cooperation and coordination across diverse stakeholders and sectoral 'jurisdictions' (World Water Assessment Programme, 2012).

---

[1] The word "water" will consistently refer to the wider water resources sector, and not the municipal water supply and sanitation one.

However, despite the attention for KCD, managing water systems and providing water services to citizens in a sustainable and cost-efficient manner remain daunting challenges, in both well-developed economies and in developing countries. Countries with well-developed economies continuously re-arrange and improve on their sectoral institutions (Considine et al., 2009). Developing countries face more serious hurdles as they often lack the financial capacity to implement policies to achieve higher sectoral standards and to invest in infrastructure and in better management. They also miss the longer administrative experience and social capital that societies in the developed economies generally possess and use in overcoming such hurdles (Franks, 1999; Therkildsen, 2000; Yusoff, 2005). At water sector level, KCD has not been analysed frequently, perhaps because of its complexity.

1.4   ASSESSING THE EFFECTIVENESS OF KCD INITIATIVES

As in the development process, there are many unknowns in KCD. Often, KCD initiatives are embedded in other programs and are not tracked separately, making it difficult to evaluate the efficacy of interventions. And even if they were monitored closely, KCD is a long-term process, subject to many external factors. The development of knowledge and capacity cannot be easily attributed to one intervention or even to a particular donor programme (Whyte, 2004). However, the methodological difficulty in determining an accurate correlation between investment in capacity and improved sectoral performance is no proof that there is no return (United Nations, 2006).

To improve the effectiveness of KCD programs and interventions, a deeper understanding of the dynamic process of capacity and knowledge development and the numerous contextual factors that influence capacity, is required. Conventional capacity assessments focus on taking snapshots of identified constituent elements, as summarised by Baser and Morgan (2008) and Brown et al. (2001), effectively ignoring the dynamics of the process and the ongoing influence of the context. Moreover, the scientific research on the foundations of KCD and its constituent elements is itself remarkably limited (Lusthaus et al., 1999; Polidano, 2000; Brown et al., 2001; Bateson et al., 2008), specifically with reference to the water sector (Blokland et al., 2009; Tropp, 2007; Cleaver et al., 2005; Biswas, 1996). The paucity of research on the use of analytical tools and frameworks to measure progress over time in KCD, has been identified by Pascual Sanz et al. (2011), Alaerts and Kaspersma (2009), Baser and Morgan (2008) and Mizrahi (2004). A difference is observed between academic approaches to measuring effectiveness and approaches from practice such as developed by donors. Practice based approaches often use results-based management tools. With such approaches it is difficult to see the longer-term effects, but when capacities can be defined unambiguously and specifically, and when it is relatively straightforward to define indicators, results-based management may be appropriate (Zinke, 2006). I will discuss this in further detail in the Theory Chapter.

The European Centre for Development Policy Management (ECDPM) has made considerable progress with the introduction of the 5 Core Capabilities (5Cs) approach (Baser and Morgan, 2008; Fowler and Ubels, 2010). This represents one of the most comprehensive studies of the concept of capacity, and the 5Cs approach is currently being tested in a number of agencies in the Netherlands. However, the core capabilities need to be described by a set of indicators that are distinct to each of the core capabilities. Furthermore, elements of capacity that are more intangible, such as culture, interpersonal relations, power and personality require more attention (IOB, 2011). In short, the application of the 5Cs approach in practice has been accompanied by difficulties.

In earlier work, Alaerts and Kaspersma (2009) introduced a conceptual model more specifically geared to the water sector (but in principle also applicable in other environments) that draws together the different elements of KCD. The conceptual model distinguishes three levels, i.e. the institutional level, organisational level and individual level, and specifies in broad terms, for each of the three levels of action the sequence of what knowledge and capacity imply, the means by which the knowledge and capacity development can take place, what the outcomes of KCD are and how these can potentially be assessed. The levels are "nested', that is, individuals operate within their organisational context, and the organisation operates within its broader context. In Chapter 2 where I explain the theoretical basis of this thesis more elaborately, I will assess this model and other approaches. In the subsequent case studies, I will study whether this framework is a useful tool that can provide reliable insight.

1.5   INTRODUCING THE AIM AND APPROACH OF THIS THESIS

An adapted version of the Alaerts and Kaspersma (2009) KCD conceptual model serves as a basis, and ordering framework, in the investigation of KCD in the water sector. In this thesis I aim to answer the following research questions:

1. How does the institutional environment influence the development and use of knowledge in the public water sector?
2. How does the organisational structure influence the development and use of knowledge in the public water sector?
3. What KCD mechanisms are available at the broader institutional, organisational and the individual levels, and to what extent are they used?

In addition I will draw upon theory from the fields of human resource development, learning, organisation and management sciences and policy analysis to explain the relations between different components of the KCD system.

At the individual level I adopt theory on professional competence (Cheetham and Chivers, 2005; Sultana, 2009; Oskam, 2009) to help understand the composition of knowledge and capacity at the individual level and the combination of different competences required by water professionals (Chapter 6). At the level of the organisation, I use Burns and Stalker's classification of mechanistic and organic organisational structure (1961) and Mintzberg's structure in fives theory (1980) to explore how formal organisational structure influences KCD (Chapter 5). At the level of the institutions I draw upon theory on advocacy coalitions (Sabatier and Jenkins-Smith, 1993) and the multiple streams framework developed by Kingdon (1995) to explain how coalitions continuously need to promote their agendas, which embody new knowledge and capacities, in order to influence existing policy regimes supported by the establishment. A window of opportunity (Kingdon, 1995), often brought about by external events that trigger a political reaction, is required to change to a new paradigm, which allows/demands the inclusion of new knowledge and capacities (Chapter 4).

I apply the adapted framework and additional supportive theory to study KCD in two public sector organisations that are representative for the water sector, or have been representative for the sector for a long time, in their institutional context. Government organisations have important development tasks and at the same time often experience difficulties in acquiring and maintaining an adequate level of knowledge and capacity in the organisation, which may yield interesting research material. I further confine myself to the water resources sector, thus excluding the water supply and wastewater utilities, to make the analysis slightly less broad.

The Directorate General of Water Resources (DGWR) of the Ministry of Public Works (MPW) of Indonesia was chosen as a case, because of the availability of a large body of accessible information regarding the development that the country's water sector has gone through in the recent past, and because the DGWR is by far the largest employer of water professionals in Indonesia. Much written information is available online, and additional information is available because of a long standing relationship between the Unesco-IHE Institute of Water Education (Unesco-IHE) and the DGWR. As a Unesco-IHE employee I had good access to this information. In the Indonesian case I have chosen to assess KCD by starting the analysis through one specific KCD intervention, namely International Post-graduate Education (IPE). In many developing countries and countries in transition, IPE is an assumedly important means of accessing global knowledge that is assumed not available locally. Annually, the DGWR sends a significant number of young professionals to universities abroad, mostly to universities in Australia, the United States and the Netherlands. It could well be that IPE is relatively more important as a means for KCD in countries such as Indonesia, as alternative means of developing knowledge and capacity are not always available, such as extensive professional networks, an organisational culture that allows criticism and self-reflection and collaborative learning with stakeholders.

The second case is the executive arm of the Dutch Ministry of Infrastructure and Environment, the Rijkswaterstaat. This case was chosen, first because of the availability of a large body of information regarding the development that the water sector has gone through since the 1950s. Second, because I wished to investigate how knowledge and capacity develop and influence decision making in an organisational unit similar to the DGWR, but located in a relatively well developed economy where I hypothesise that a wide array of KCD mechanisms is available to generate and exchange new knowledge. Because comprehensive background review studies are available for the Dutch case, and not for the Indonesian case, the Indonesian case study requires more in-depth study. Additionally the Indonesian case is more challenging because a foreign researcher may face cultural barriers in obtaining critical opinions and information. I discuss my approach to this challenge in Chapter 3.

I adopt a mixed method approach using surveys and semi-structured interviews in both cases to analyse how water professionals acquire knowledge and capacities, and I undertake a historical analysis of both the Indonesian water sector and the Dutch water sector to study how the environmental, and to some extent the cultural, features and priorities in society at distinct junctures in time have influenced the use of certain KCD mechanisms.

1.6   STRUCTURE OF THE THESIS

The thesis is structured to align with the conceptual model that I introduce in Chapter 2 and subsequently.

The manuscript consists of in total eight chapters, five of which are presenting the results of particular research segments:

- Chapter 2 presents a review and discussion on KCD
- Chapter 3 explains the selection of the cases, provides a research strategy and the methods,
- Chapter 4 is the first of the chapters on the Indonesian case study and explains the role of KCD at the institutional level in the DGWR,
- Chapter 5 explains the organisational structure of the DGWR and its influence on KCD,
- Chapter 6 is the last of the chapters on the Indonesian case study and focuses on individual KCD in the DGWR, and determines the impact of IPE on individual KCD,
- Chapter 7, in which the KCD conceptual model is applied to the case study of the Rijkswaterstaat,
- Chapter 8 links the different research segments, discusses the results and draws conclusions on scientific and development relevance.

2 KCD in public water management: an initial conceptual model

2.1 INTRODUCTION[2]

It is nowadays generally agreed that capacity enhancement involves more than the strengthening of individual skills and abilities. Trained individuals need an appropriate environment, and the proper mix of opportunities and incentives to apply their acquired knowledge. Understanding capacity development therefore requires a more comprehensive analytical framework that takes into account the individual, the organisational and the institutional levels of analysis (Alaerts and Kaspersma, 2009; Lopes and Theisohn, 2003; Morgan, 1993). In this chapter I provide an overview of KCD research to date, I unpack the different components that make up KCD, and I discuss several conceptual models of KCD. I build further on a conceptual model developed earlier by Alaerts and Kaspersma (2009). In addition, I allude to the complementary theory that is required to deepen the understanding of capacity development in and between the individual, organisational and institutional levels.

2.2 DISTILLING THE THEORETICAL CONCEPTS

2.2.1 Capacity

Different authors define capacity according to the theoretical (or political) frameworks from which they approach the subject. This means that there is no completely common capacity language or set of terms yet to help communication and shared learning on capacity (Ubels, 2010; Baser and Morgan, 2008; Morgan, 2006; Brown et al., 2001; Brinkerhoff, 2005). In 1993 Morgan defined capacity as 'the ability of individuals, groups, institutions and organisations to identify and solve development problems over time' (1993). The European Centre for Development Policy Management (ECDPM) defined capacity as the overall ability of an organisation or system to create public value, focussing on the level of the organisation.

Hildebrand and Grindle (1994) focused on the dynamic aspects of capacity, stating 'Capacity is the ability to perform appropriate tasks effectively, efficiently, and sustainably. This implies that capacity is not a passive state - the extent of human resource development, for example - but part of an active process'. Alaerts (1999) later amalgamated these concepts in his definition of capacity for the water sector, stating: 'Capacity allows individuals, organisations and relevant institutions to perform in such a way that the sector as an aggregate can perform optimally, now as well as in the future'.

In positing their definitions all the authors mention that capacity is about the ability to do 'something' successfully, within a specific context over time. The particular ability differs from one context to another, and may change with time. In the context of organisations the required abilities may be problem solving, managing affairs or executing tasks to contribute to better performance (Lusthaus et al., 2002). Capacity development corresponds to the goal of people wanting to learn these abilities and increase their

---

[2] Part of this chapter has been published earlier in Alaerts and Kaspersma Alaerts, G. J., and Kaspersma, J. M.: Progress and challenges in knowledge and capacity development, in: Capacity Development for improved water management, edited by: Blokland, M. W., Alaerts, G. J., Kaspersma, J. M., and Hare, M., Taylor and Francis, Delft, 327, 2009.

options and choices. This applies in a similar fashion to organisations, institutions and societies as a whole (Lopes and Theisohn, 2003), as affirmed in the quotes by two of the authors above who indicate that capacity is a property possessed at different levels – individual, organisational and institutional, that may be expressed in an aggregate fashion. In this thesis, I adopt these different levels of action and approach capacity and its development from a systems perspective, meaning that capacity cannot be explained in isolation of its surrounding context and that each level is related to and dependent on capacity at the other levels.

Other authors, most notably Baser and Morgan (2008), use the term 'capability', when they talk about capacity at organisational or systems level. Given my definition of capacity, which includes all levels of action, i.e., individual, organisational and the broader institutional level, I do not consider it useful to adopt distinct terms for capacity at the individual level, and at the organisational level and higher. Moreover, by using the term capacity, I avoid confusion with the work of Sen (1999) and ECDPM (Baser and Morgan, 2008) as discussed in Section 2.3.2 and 2.3.5.

2.2.2 Knowledge

A large proportion of the literature on knowledge management is geared to corporate businesses and firms, e.g. Sveiby, and Nonaka and Takeuchi (1995). For the public sector, the rationale for knowledge management is just as important. Instead of the maximisation of profit under conditions of competition, public service delivery is required to be maximized at minimal cost, under pressure from a society that demands a healthy public sector and a low tax burden (Alaerts and Kaspersma, 2009).

According to Weggeman (1997) knowledge 'is the personal capability that enables an individual to execute a certain task'. This relates to the vision of Nonaka and Takeuchi, who emphasize: 'knowledge is essentially related to human action' (1995). Sveiby (2001a) equates knowledge to the active capacity of an individual. Each individual has to re-create their own capacity to act and their own reality through experience. This definition coheres with a more detailed definition of Weggeman (1997), namely: 'Knowledge is linked to capacity in the sense that knowledge is the product of information, and the capacity to act on this information, through experience, skills and attitude. This process will lead to a result, and the appreciation of that result is dependent on the individual judging it.'

Epistemological analysis reveals that knowledge can be both explicit, referring to knowledge in a form independent of the originator and therefore accessible by others (e.g. books, models, tools etc.), and tacit, referring to knowledge embedded within a person (such as the ability to ride a bicycle) (Tsoukas, 2002; Sveiby, 2001b; Nonaka, 1994; Polanyi, 1966). Polanyi (1966) differentiates tacit from explicit knowledge, saying 'we know more than we can tell' and meaning that explicit knowledge is the knowledge that we can tell, while tacit knowledge is what we know but find hard to tell. Weggeman (1997) agrees with this conceptualisation, stating that 'Tacit knowledge is personal knowledge that is difficult to formalize and therefore difficult to share with others'. It encompasses experience, skills and attitude and has a technical and a cognitive dimension. The technical dimension includes implicit know-how such as skills and craftsmanship, nurtured by years of experience, and exemplified in the craftsman's difficulty in pointing to the scientific and technological basis of his skills. The cognitive dimension includes mental models, values, beliefs and assumptions. These are so deeply rooted that they have become self-evident. The cognitive dimension determines the way in which an individual perceives the world around them. In distinguishing four different types of knowledge for which different methods of learning are appropriate, Alaerts and Kaspersma (2009) build upon the Weggeman conceptualisation. They also provide illustrations from the water sector of the types of knowledge concerned. First they distinguish information, or factual knowledge illustrated as 'water boils at 100°C'. Second, they distinguish understanding (through experience), which can be illustrated by the phrase 'why does it rain?' Third, they distinguish skills such as language proficiency or the ability to work in a team. Fourth, they distinguish attitudes such as problem-solving attitude, the capability to approach a complex challenge, ambition, 'gut feeling', and the drive to keep learning. In this paper, we adopt the conceptualisation of Alaerts and Kaspersma (2009).

2.2.3 Learning

There are many schools of thought on learning relevant to research on knowledge and capacity development, their only agreement being the assumption that learning entails a future improvement in performance (Fiol and Lyles, 1985). Their different standpoints are summarized briefly hereafter.

Behaviourism views the process of learning as a mechanism - the result of a behavioural response to some form of stimulus, such as in the experiments of Pavlov (1927). If a particular response repeatedly results in a reward and/or reinforcement, then learning can be expected to take place (Cheetham and Chivers, 2005). Behaviourism has a positivist approach and has been criticized for largely ignoring the influence of thoughts, feelings and attitudes in learning processes. However, aspects of this learning theory remain relevant, and feedback and appropriate reinforcement are still considered important in education. Where behaviourism is based on a positivist approach, suggesting that input A will lead to output B with the appropriate reward or reinforcement, cognitive learning is based on a constructivist approach, and concerned with what goes on between input A and output B, the mental processes such as reasoning or problem solving.

Secondly, cognitive approaches look at the way people absorb information from their environment, sort it mentally and apply it in everyday activities. Cognitive learning is closer to the concept of knowledge creation and differs from behaviourism in that it takes into account the implicit component of knowledge, accommodating the attitude of the learner.

Experiential learning, however, 'rests on a different philosophical and epistemological base from behaviourist theories of learning and idealist educational approaches', according to Kolb (1984). Experiential learning assumes that ideas are formed and reformed through experience. 'Learning is the process whereby knowledge is created through the transformation of experience'. Kolb literally states that learning is knowledge creation, emphasising particularly the use of the experience of the learner, instead of the attitude of the learner as in cognitive approaches.

In this dissertation I understand learning to be synonymous with knowledge creation, because learning takes into account the same components as knowledge creation, i.e. learning is the combination of information or factual knowledge, with understanding, skills and attitude. It is not only experience, as in experiential learning, nor only attitude, as in cognitive learning, but, as in knowledge creation, a combination of explicit knowledge (factual knowledge or information, and understanding) and all tacit forms of knowledge: experience, attitude, and existing skills.

The understanding that learning is akin to knowledge creation can be applied to the different levels at which knowledge is created e.g. to the individual and organisational levels. Education is an important but not the exclusive means of creating knowledge at the individual level.

2.2.4   Individual vs. organisational learning

The importance of individual learning for organisational learning is at once obvious and subtle, according to Kim (1993). Obvious, because all organisations are composed of individuals; subtle, because organisations can learn independently of any specific individual, but not independently of all individuals. Organisational learning is not simply the sum of individual learning (Fiol and Lyles, 1985; Kim, 1993; Levitt and March, 1988; Liao et al., 2008), because organisations that value learning at the organisational level, will ensure that individuals learn, but moreover ensure that an appropriate infrastructure is in place for knowledge and information exchange, that working in groups is encouraged and that incentives are in place for knowledge sharing (Argyris, 1993; Kim, 1993; Fiol and Lyles, 1985). Organisational learning can only be successful when it is based on an understanding of how the whole organisational system is connected, rather than focusing on individual parts (Senge, 1990).

2.2.5   Complexity

Further, more recent work (Ramalingam et al., 2008; Land et al., 2009; Woodhill, 2010) emphasizes the complex and systems nature of capacity development efforts and argues that the capacity of an organisation is both a distinct entity in itself as well as the result of the capacities of the individuals in the organisation. It is the consequence of a wide variety of inputs and attributes such as the types of knowledge that have been transferred, the organisational structure and procedures, the leadership and managerial capabilities of the individuals, amongst others. All these attributes tend to change over time and mutually influence one other. This complexity blurs the relationship between the capacity development input, and its outcome.

2.3   CONCEPTUAL MODELS TO EXPLAIN OR ASSESS KCD

2.3.1   Overview of KCD assessment models

Scientific research on the foundations of KCD and its constituent elements is only starting to emerge. For some examples see Belda (2012), Oswald (2010), Lusthaus (2002), Polidano (2000), Brown et al. (2001) and Bateson (2008), and specifically with reference to the water sector: Gupta et al (2010), Blokland et al. (2009), Tropp (2007), Cleaver et al. (2005) and Biswas (1996). Non-academic approaches are for example Outcome Mapping (Earl et al., 2001), Most Significant Change Theory (Davies and Dart, 2007), and the five Core Capabilities Approach (5Cs), each based on a complexity approach (Section 2.2.5). The Learning Network on Capacity Development (LenCD) has produced a comprehensive and well structured overview of models and organisations working on capacity development (Pearson, 2011).

Most conventional capacity assessments focus on taking snapshots of identified constituent elements, as summarised by Baser and Morgan (2008) and Brown et al. (2001), effectively ignoring the dynamics of the process and the ongoing influence of the context. The paucity of research on the use of analytical tools and frameworks to measure progress over time in KCD, has been identified by Pascual Sanz et al. (2011), Alaerts and Kaspersma (2009), Baser and Morgan (2008) and Mizrahi (2004). Clearly, assessing capacity development continues to present both a scientific and a development challenge. Table 2.1 presents a snapshot of some of the current models, their theoretical foundations and approach to understand and assess capacity development.

Table 2.1.: Academic and non-academic methods and approaches to understand and assess capacity (adapted from Pascual Sanz (forthcoming)).

Name of the approach	Type of assessment	Reference
Institutional assessment and Capacity Development - European Commission	Institutional capacity	(EuropeAid, 2005)
Adaptive capacity wheel	Institutional capacity	(Gupta et al., 2010)
Outcome Mapping	Change in behaviour of an organisation/organisational capacity	(Earl et al., 2001)
Most significant change - Story telling	Individual and organisational capacity and change	(Davies and Dart, 2007)
5Cs framework	Organisational capacity	(Baser and Morgan, 2008)
Framework for organisational assessment	Organisational capacity	(Lusthaus et al., 2002)
UNDP approach to measuring capacity	Results oriented	(UNDP, 2010)
WBI Capacity development and results framework (CDRF)	Results oriented	(Otoo et al., 2009)
Human Capabilities Approach	Individual capacity	(Kuklys, 2005; Sen, 1999; Robeyns, 2005)

The capacity assessment systems of development organisations such as the United Nations, the Asian Development Bank (ADB) and the World Bank (WB) focus on measuring tangible elements in the activity-output-outcome-impact chain (ADB, 2006; Otoo et al., 2009; UNDP, 2010). Examples are Results Based Management (RBM) used by the ADB and the Capacity Development Results Framework (CDRF) used by the World Bank. However, the vagaries of the capacity development process and the influences of external or unknown factors are not accommodated well, leading to workable but flawed assessment models.

A number of approaches to capacity development deserve attention, as they have attracted great interest in academic and/or development communities: the Human Capabilities Approach (HCA) advocated by Amartya Sen (Robeyns, 2005; Sen, 1999), the adaptive capacity wheel (Gupta et al., 2010), the framework for organisational assessment (Lusthaus et al., 2002), the 5Cs approach of ECDPM (Keijzer et al., 2011; Baser and Morgan, 2008), and the KCD conceptual model by Alaerts and Kaspersma (2009).

2.3.2 Human Capabilities Approach

According to Sen (1999) fundamental to enlarging the choices of people is the development of human capabilities — an individual's actual and potential activities and states of being respectively, and functionings – the various things a person may value being and doing, such as being in good health, having sufficient food, having an education (Kuklys, 2005; Comim et al., 2008; Deneulin et al., 2006). A capability represents the various combinations of functionings. The most basic capabilities for human development are to lead long and healthy lives, to be knowledgeable, to have access to the resources needed for a decent standard of living and to be able to participate in the life of the community. Without these, many choices are simply not available, and many opportunities in life remain inaccessible (UNDP, 2008b). A point of criticism in the debate about the approach is a lack of clarity about the relationship between the individual and society in Sen's writing about capability (Comim et al., 2008). However, social structures and institutions have an important effect on people's capability sets and in that sense they are implicitly included. Social structures and institutions are included in the conceptual framework of the capability approach, although with the clear recognition that these are the means and not the ends of well-being (Robeyns, 2000). The approach does not consider capabilities at the level of groups or the level of institutions, which is also a natural consequence of the way capability is defined in Sen's approach, and in that respect it differs from my chosen approach. I argue that capability, or rather capacity at the level of an organisation is not just the sum of individual capabilities, but that the synthesis of individual capabilities can yield additional value if the organisation is open to learning and, indeed, if the environment is 'enabling'.

2.3.3 The Adaptive Capacity Wheel

This approach to adaptive capacity originates in the study of adaptation to climate change. Because climate change brings unpredictable changes, it calls for institutions that enhance the adaptive capacity of society (Gupta et al., 2010). The authors indicate that little research is available on assessing institutions with regard to their ability to enhance the adaptive capacity of society. However, the authors use the term 'institutional adaptive capacity' and the 'institutional capacity to enhance the adaptive capacity of society' interchangeably, which in my opinion acts to confuse rather than clarify. I argue that these are two distinct processes.

The proposed framework to assess institutions is well worked out and tested and has delivered the following story line[3]: institutions that promote adaptive capacity are those institutions that (1) encourage the involvement of a variety of perspectives, actors and solutions; (2) enable social actors to continuously learn and improve their institutions; (3) allow and motivate social actors to adjust their behaviour; (4) can mobilize leadership qualities; (5) can mobilize resources for implementing adaptation measures; and (6) support principles of fair governance. These six dimensions of adaptive capacity have twenty-two criteria. The model is one of the few that pays explicit attention to institutional learning and furthermore assigns a distinct role to leadership.

Although the 6 dimensions and associated criteria are well defined, I prefer to work with a model that gives full consideration to the capacity and responsibilities at institutional, organisational and individual level distinctly and simultaneously.

2.3.4 Organisational assessment framework

The framework by Lusthaus et al (1999) is appropriate for the characterisation of organisational capacity, and is less helpful at the levels of the sectoral institutions and the individual. It suggests examining eight interrelated areas to evaluate capacity: strategic leadership, financial management, organisational structure, organisational infrastructure, human resources, program and service management, process management and inter-organisational linkages (Pascual Sanz et al., 2011). The critique of Sanz et al. is that Lusthaus et al. follow a mechanistic approach, where capacity is the sum of the parts mentioned earlier and all parts can be analysed. In my view the institutional environment and individual capacity play a significant role, not only in determining organisational performance, which Lusthaus et al. concede, but also in the assessment of organisational capacity.

2.3.5 5 Capabilities approach

ECDPM has introduced the 5Cs approach. (Baser and Morgan, 2008; Fowler and Ubels, 2010). The 5Cs stand for five core capabilities that can be found to a greater or lesser extent, in all organisations or systems: the capability to act, the capability to generate development results, the capability to relate, the capability to adapt and the capability to integrate (Fowler and Ubels, 2010).

The framework is one of the more comprehensive studies of the concept of capability, and the 5Cs approach is currently being tested in a number of agencies in the Netherlands. At the basis of the model lies a considerable amount of practical experience that reveals the complex nature of capacity (Ubels, 2010). ECDPM defines a capability as the collective skill or aptitude of an organisation or system to carry out a particular function or process either inside or outside the system. Capabilities enable an organisation to do things and to sustain itself (Baser and Morgan, 2008). In contrast to the Human Capabilities Approach, which focuses on the individual level the 5Cs approach has the organisation as its focal point. The case studies so far have represented small

---

[3] Literally taken from Gupta et al. Gupta, J., Termeer, C., Klostermann, J., Meijerink, S., van den Brink, M., Jong, P., Nooteboom, S., and Bergsma, E.: The Adaptive Capacity Wheel: A method to assess the inherent characteristics of institutions to enable the adaptive capacity of society, Environmental Science and Policy, 13, 459-471, 10.1016/j.envsci.2010.05.006, 2010.

organisations up to national systems (IOB, 2011), but always with the organisation at the centre of attention. The interactions between the 5Cs differ in each situation, according to Fowler and Ubels (2010), and need to be operationalised for every situation in reality. Critique, mainly ventilated in a study by the Policy and Operations Evaluation Department (IOB) of the Dutch Ministry of Foreign Affairs (2011) indicates that the 5Cs need to be described in less abstract terms in order to explain their functions and significance. The core capabilities may need to be described by a set of indicators that are distinct for each of the core capabilities. Furthermore, elements of capacity that are more intangible, such as culture, interpersonal relations, power and personality require more attention. In addition, more robust methods are required to help organisations identify organisation-specific indicators to assess the development of capacity. The application of the approach in reality is therefore not without difficulty. However, the current testing in a variety of cases may yield interesting results that help to refine the model further.

2.3.6   The KCD conceptual model

As individuals are nested in organisations and organisations in turn are nested in their broader institutional contexts, I consider the development of capacity at each level to be dependent upon the capacity at the other levels and view the conceptual frameworks that do not take this into account as insufficiently comprehensive. I explicitly choose to work with a conceptual model that allows any level of analysis as an entry point, be it the individual, organisational or institutional level. From the chosen perspective, it should be possible to indicate what a certain measure or intervention implies for capacity development at the other levels. The conceptual model of Alaerts and Kaspersma (2009) allows for an entry point to assess capacity at every level, in contrast to both the HCA and the 5Cs frameworks, not only indicating that the other levels need to be conducive for capacity, but providing the option to assess capacity at these levels. Rather than confusing Institutional capacity with the extent to which institutions are conducive for capacity at a lower level, the conceptual model of Alaerts and Kaspersma (2009) focuses on capacity at each level and can be used as an ordering framework to focus investigation at each level. The model does not encompass all the requisite theory to explain how knowledge and capacity develop in a certain context and at a certain level. Necessary complementary theory is drawn upon and explained for each level separately in later chapters.

2.4   THE ADAPTED KCD CONCEPTUAL MODEL

In this thesis I use an adapted version of the KCD conceptual model presented in earlier work by Alaerts and Kaspersma (2009) (Figure 2.1). A number of nested levels is distinguished, that is, the individual is embedded within their organisation, and the organisation operates within its broader environment. The broader context or enabling environment is itself divided into the part that typically falls within the realm of the formal institutional arrangements, and into a second part which accommodates the societal context. (Civil) society forms part of this enabling environment and at the same time is an actor in its own right, as within society numerous formal and informal non-governmental networks, associations and organisations take part in the broader game of

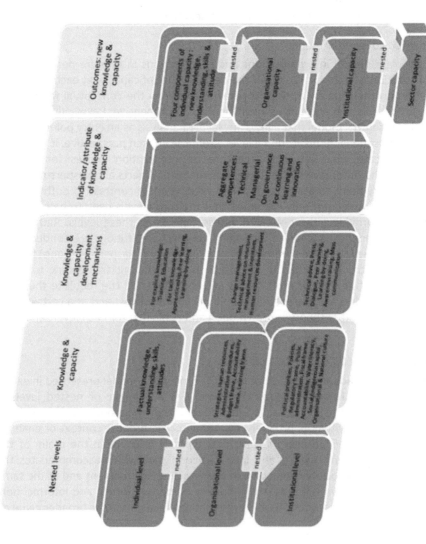

Figure 2.1. KCD conceptual model (adapted from Alaerts and Kaspersma (2009)

38

water management nationally and internationally (Madani, 2010; Slinger et al., 2010; Teasley and McKinney, 2010).

For each of the three levels, the individual, the organisation and the enabling environment (column 1), the sequence of (i) what knowledge and capacity imply (column 2) (ii) the means and mechanisms by which the knowledge and capacity development can occur (column 3), (iii) the indicators or attributes by which these can be assessed (column 4), and (iv) the outcomes (column 5), are specified in the diagram (Figure 2.1). In the initial model column 4 and 5 were reversed and the outcome column changed from indicating organisational performance, good governance and sector performance to indicating capacity instead. The choice to specify the outcomes in column 5 of the revised conceptual model in terms of individual, organisational and broader institutional capacities, together comprising the sector capacity, reflects the insight that a comprehensive assessment of KCD must necessarily address capacity at each of the levels of a system simultaneously.

The adapted conceptual model is designed for a water resources management context and the identified aggregate competences are applicable to this context. The model is nevertheless considered sufficiently generic to be applied within other contexts such as environmental management, yet specific enough to be operationalised in a variety of contexts. This is however not tested in this thesis.

As reflected in the upper building blocks of the adapted KCD conceptual framework (Figure 2.1) the performance of the water sector derives from the effective action of individuals with appropriate knowledge and capacity, who function in larger organisations (such as ministries, local governments, water user associations, civil society organisations, etc.), as represented in the middle building blocks. The effectiveness of these organisations depends both on the effectiveness of the individuals and on the typical features that shape the capacity of the organisation itself through its skills mix, its internal operational and administrative procedures. However, even organisations with suitable professionals and appropriate procedures still need an enabling environment to put in place the facilitating institutional factors such an enabling legal and regulatory framework, financing and fiscal rules that stimulate action, expressed in policies, and a broadly supportive political climate in the parliaments and among the voters and consumers. The enabling environment is represented in the lowest row of building blocks in Figure 2.1, together with the civil society, which can also be seen as part of the institutional environment. KCD has been identified as essential to support and implement the improvement of institutions or changes in institutional arrangements (Franks, 1999; Alaerts, 2009a). Often it is difficult to distinguish KCD proper from institutional development, and, indeed, KCD is embedded in and in effect helps shape any institutional development and reform effort. Accordingly, KCD affects its enabling environment and is part and parcel of change management (Morgan et al., 2010; Mugisha and Brown, 2010).

Many instruments for knowledge transfer and/or knowledge creation are included within the adapted KCD conceptual model and are represented in column 2, for each level. At the individual level knowledge, understanding and skills are generally developed through typical knowledge transfer instruments such as education and training, however, whether the desired knowledge (or capacity) is explicit or tacit makes a difference to the choice of instrument. As Polanyi (1966) and Sveiby (2001b) argued from their respective epistemological and practical management insights, tacit knowledge is eventually far more important as it shapes skills and deeper attitudes. Tacit knowledge can best be transferred through one-on-one interaction between junior and senior, apprentice and teacher. Organisational capacity development is achieved by educating or training the (staff) members and by helping the organisation as such learn from the experiences of others. Technical assistance, management advice, learning experiences, comparison with peers and benchmarking, are important instruments. Pilot projects provide a confined field setting to learn about specific innovations (Vreugdenhil et al., 2010). Networks play an increasingly important role in generating, sharing, corroborating and improving knowledge and capacity both for individuals and for organisations. Indeed, networks - both formal and informal associations and 'communities of practice' - are becoming the main mechanisms for professional improvement for many water professionals (Luzi et al., 2008). In addition, Information and Communication Systems (ICS), such as online communities and social media, are powerful tools to support and intensify communication and open up new avenues for the dissemination of knowledge, including best practices. At the level of the enabling environment, governments and other actors also 'learn' and acquire the capability to become more enabling. Policy makers, governmental departments and politicians, can increase their understanding about new challenges and solutions by, for example, drawing lessons from international 'good practice'. Technical advice, communication platforms and peer-learning are useful instruments for exchanges of information. In civil society, capacity already exists in the form of what is often called social capital or indigenous knowledge that resides in communities, and in water literacy. This can be further developed through comparison activities and peer-learning with other communities. The press and mass communication-such as Discovery Channel - are powerful vehicles to disseminate knowledge and develop attitudes, and reach many communities that otherwise receive little opportunity to communicate with the government. Non-governmental organisations and networks are important and often effective actors in this capacity development. Finally, the role of society is of course critically important as it shapes the nation's consensus about the priorities for the future (and the budget allocations) through electing its political representatives and holding its government accountable.

The adapted KCD conceptual model is completed with a conceptualisation of competence (Table 2.2). By positing that capable individuals, the organisations in which they work and the institutions in which both are embedded possess aggregate competences to act and learn, four aggregate competences are distinguished (column 3). First, technical competence is required to analyse and solve the problems which are of a technical nature. The technical competence can be subdivided into the different components of knowledge distinguished in Chapter 6, Section 6.2. In this light, technical competence consists of factual knowledge or information of a technical nature, the understanding of that technical knowledge including its application, and moreover having the appropriate skills to work with the knowledge, e.g. being able to work in a team, drilling a hole, or writing a report; in addition, acquiring a problem-solving attitude and a drive to keep learning. For instance, all sector agencies need to regularly acquire new technical knowledge on an array of subjects from piling and construction techniques to climate change mitigation and adaptation. To possess an aggregate technical competence, all four levels of knowledge must be present. The concept of competence and its relation to knowledge will be further investigated in Chapter 6.

Second, organisations need to have an adequate pool of management competences embodied in their senior staff. In many developing countries sector agencies may score well on technical and civil engineering aspects, but often the competence to manage personnel and organisations, as well as the water resource itself, is modest.

Third, an effective and performing water sector requires organisations that possess skills to foster and apply principles of good governance, such as dialogue and communication with stakeholders, resource allocation within policy models that aim for equity and poverty alleviation, transparency and accountability.

Finally, as mentioned before, capable individuals and organisations are those that manage, by deliberate decision, to innovate and to keep learning. Learning and innovation do not come automatically but require financial resources and personal and managerial procedures to foster knowledge generation and sharing. This learning can emerge from an acquired attitude, or as natural inquisitiveness, but institutions may become interested in and coaxed into learning only after being held accountable for poor performance. The competence for learning and innovation is termed a meta-competence because it exists beyond the other competences and enables individuals and organisations to monitor and develop the other competences (Cheetham and Chivers, 2005). The aggregate competences may differ if the framework were to be applied in another context, although for the public sector some of the aggregate management and governance competences would remain valid. For instance, the meta-competence for continuous learning and innovation is valid in any context.

Table 2.2. Illustrations of the four aggregate competences at the individual, organisational and institutional levels for use in the water sector

	Individual level	Organisational level	Institutional level
Technical competence	Regularly updated factual knowledge and skills, Understanding of the broader technical context.	Appropriate knowledge and skills mixes for the services that are delivered, such as engineering, legal, financial, institutional knowledge, Knowledge on procurement and investment procedures.	Technical expertise and available skills mixes in a broader setting, Systems for critical review and corroboration of knowledge and information.
Management competence	Project management skills, Financial management skills, Personnel and team management skills, Mentoring skills, Ability to 'deliver', Leadership. Mentoring skills, Ability to 'deliver', Leadership.	Leaders able to operate with goals and objectives as agreed with supervisory entities and main stakeholders. Ability to set goals, strategy, Financial management, People management, Appropriate staff rotation; talent spotting, incentive systems, Project management, Ability to 'deliver' timely.	Sound and workable task assignments of sector agencies, Minimal overlap between agencies, and size and task of agencies facilitate proper management and task execution, Sound financial, fiscal and budgeting systems, Facilitating proper management by organisations.

Governance competence	Understanding of procedures, Understanding of political consensus building, Ability to engage with and listen to stakeholders, Ability to apply inclusiveness, Focus on results.	Transparent decision making processes, Procedures to consult with stakeholders, and provide empowerment to others, Procedures to be held accountable, including transparency in budgets and plans.	Distinction between 'operator' and 'regulator', Policy to ensure inclusiveness in particular regarding objectives, priorities and strategies, Policy to ensure transparency and accountability.
Competence for continuous learning and innovation	Desire to 'keep learning', readiness to critically reflect on one's own performance, Availability for training and education in new skills and knowledge.	Readiness and procedures to critically review organisations performance on a continuous basis, and revise if necessary, Goal, procedures and resources to support learning by staff, organisation and if necessary other stakeholders, Support of 'communities of practice', and rewards for staff learning.	Policy to promote open working atmosphere and critical reflection on performance, openness to review sector performance on a continuous basis, and revise policies and arrangements if necessary, Foster inclusiveness.

2.5 COMPLEMENTARY THEORY

The adapted KCD conceptual model does not clarify the relations between existing knowledge and capacities and their development at each level of the model or between the levels. Complementary theory is required to operationalise the relations between the various levels and to explain how KCD works at each level. At the institutional level, Kingdon's Multiple Streams Framework is adopted (1995) to explain how knowledge and capacity are selectively used in policy making processes. This is explained in Chapter 4 and utilised again in Chapter 7. At the organisational level, I draw upon Mintzberg's structure in fives (1980, 1979) to characterise the formal organisational structure, and Burns and Stalker's theory on organic and mechanistic organisational structure (1961) to indicate the degree of openness of organisations to new knowledge and capacities, discussed in Chapter 5 for the Indonesian case study and in Chapter 7 for the Dutch case study. At the level of the individual, I adopt part of the model of professional competence developed by Cheetham and Chivers (2005, 1996) and Oskam's T-shaped competence profile (2009), further explained in Chapter 6.

2.6 CONCLUSIONS

This chapter has articulated the concepts that form the core of KCD and has consistently applied these concepts in the formulation of a conceptual model of KCD.

Scientific research on the foundations of KCD and its constituent elements is limited, especially with reference to the water sector.

Other conceptual models were investigated but were discarded for use in this thesis because they either focus on one level only, not reflecting the nested complexity of the real system, or because they have proved difficult to use in practice. The reviewed literature has revealed that a tension exists between models that are workable and operationally applicable and models that are more complex, reflect reality more fully, but are difficult to apply in practice. The adapted conceptual model that I adopt is comprehensive in the sense that it looks at capacity at three levels simultaneously and at the interaction between these levels, but complementary theory is required to clarify the relations between knowledge and capacity and the components within the three levels.

3 Research strategy and methods

3.1 CASE SELECTION

A case study approach was adopted in applying the adapted KCD conceptual framework and complementary theory presented earlier. Case studies emphasize detailed contextual analysis of a limited number of events or conditions and their relationships. A case study is an empirical inquiry that investigates a contemporary phenomenon within its real-life context; when the boundaries between phenomenon and context are not clearly evident; and in which multiple sources of evidence are used (Yin, 2003). Because the study addresses KCD in its wider context, and it is difficult to distinguish where KCD stops and the context begins, i.e. the boundaries are blurry, this is an appropriate research approach.

I have chosen to investigate how KCD operates by varying two sets of variables, namely (a) by selecting two different water sector organisations, and (b) by studying these over longer periods of time with the intention to distinguish critical paradigmatic phases in each case. I study the different levels as identified in the adapted KCD conceptual framework.

The DGWR of the MPW of Indonesia was chosen as a first intensive case, because of (a) its track record as one of the largest professional water organisations in Asia for the past half century, and (b) the availability of a large body of study and information regarding the important development phases and transitions that the country's water sector has gone through over 40 years. In Section 2.4 I described a number of KCD mechanisms that can be adopted depending on the institutional context of the case study. With regard to the Indonesian case I hypothesise that IPE is relatively important in a society were knowledge exchange is few other KCD mechanisms are assumed to be available.

As a second, and less thoroughly analysed case, I chose the executive arm of the Ministry of Infrastructure and Environment, the Rijkswaterstaat, (a) because of the availability of a large body of study and information regarding the development that the water sector has gone through and (b) to investigate how knowledge and capacity develop and influence decision making in an organisational unit similar to the DGWR, but where it can be assumed that a wide array of KCD mechanisms are available to generate and exchange new knowledge.

3.2 RESEARCH STRATEGY

In both cases I will make a historical longitudinal analysis from approximately the 1960s until present. In the Indonesian case, the entry point in the adapted KCD conceptual model is provided by the 'post-graduate education' component, as a means of KCD for individuals. I have chosen to use this entry point because it can be hypothesised that post-graduate education is relatively more important as the use of other KCD mechanisms is limited. The use of individual knowledge and capacities is determined by the organisational and institutional context in which the alumni work after their education abroad, and in turn the organisational and institutional levels are influenced by the knowledge and capacities that enter the organisation. The resultant methodological framework, used to investigate these relationships is displayed in Figure 3.1.

The effects of two different types of post-graduate water education are assessed: international post-graduate education (IPE) and local – Indonesian - post-graduate education (LPE). The knowledge and capacities acquired during the post-graduate education are investigated by means of semi-structured interviews and a questionnaire. Furthermore other mechanisms for KCD listed in the top row of Figure 2.1 were investigated, as they may be equally or more important for KCD in organisations than post-graduate education.

As a next step in the methodology, the four competences developed and discussed in the theory chapter are assessed for individuals, and the organisation in which they work as well as the enabling environment of the organisation. The investigation focuses on obtaining an overview of the system of which post-graduation education forms a part and by which it is influenced. As a consequence, cognisance is taken of the possibility that unexpected factors or components might emerge, by asking open questions in the interviews to a broad range of respondents.

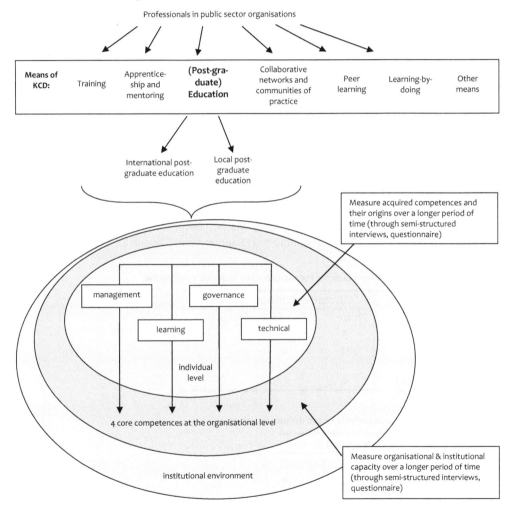

Figure 3.1. Methodological framework for Indonesian case

A similar strategy is followed in the lighter Dutch case study. In neither case does the emphasis lie on post-graduate education, but it is a methodological entry point to study the full range of KCD mechanisms.

3.3   METHODS

3.3.1   Respondent selection in the Indonesian case study

I adopted a two-pronged method for both cases, consisting of semi-structured interviews and a survey.

In Indonesia, interview respondents from the DGWR were selected to form a representative sample on the basis of their involvement in the DGWR during one or more of three development phases starting in the 1970s (defined by year of recruitment). The choice for three development phases is explained in Chapter 4. A secondary target group, for the interviews only, included people external to the DGWR, yet involved in water management in Indonesia in the same development phases, with a good overview of the water sector. Respondents from the latter group are henceforth termed resource persons. The interview and survey respondents from the DGWR were further selected based on the type of education they had enjoyed in the water field, that is, a locally-based (LPE) or international post-graduate education (IPE).

The IPE respondents acquired their Master degree at various international institutions, amongst others at Colorado State University in the USA, the University of Manitoba in Canada, (Unesco-) IHE in the Netherlands, Asian Institute of Technology in Thailand and the Indian Institute of Technology in India. The LPE respondents typically acquired a Master degree, mostly in civil engineering or water resources management, at universities such as University of Gadjah Mada, University of Indonesia, University of Sriwijaya and the Institute of Technology in Bandung and Surabaya.

Table 3.1. Interview respondents for Indonesian case in seven sub-groups

Degree type:	Started employment during:		
	Phase I	Phase II	Phase III
LPE	3	1	5
IPE	12	1	2
Resource persons	14	-	-

Table 3.2. Sample for Indonesian case divided in sub-group

	Started employment during:		
	Phase I	Phase II	Phase III
LPE	14	11	9
IPE	20	7	10

3.3.2 Qualitative data collection for the Indonesian case study

Thirty-eight semi-structured interviews were undertaken in the period from November 2008 to January 2010. Each interview lasted between 1 and 2 hours. The interviews were recorded and transcribed within a day and returned to the respondents to check the content. All interviews were anonymised.

The respondents can be categorized into seven strata as listed in Table 3.1. For the survey the sample group is categorised in Table 3.2. The number of respondents in each stratum is proportional to their occurrence in the organisation in accordance with the requirement of the stratified random sampling method.

The interviews investigate the daily tasks and activities of respondents, if applicable the contents and utility of the competences acquired during their studies, and their reflections on the competences used in the development of water management in Indonesia. Furthermore questions are asked about the enabling environment and on the use of KCD mechanisms in the organisation and organisational capacity.

In the semi-structured interviews ample opportunity is provided for respondents to bring up topics that they find important, which may not be covered by the interview questions. The interview questions can be found in Annex A.

Supplementary data was obtained of the Dominance – Influence – Steadiness Compliance (DISC) quadrant behavioural model (Marston, 1928), tested in the DGWR. 54 water professionals filled in the DISC personality test, under guidance of a resource person. I explain the theory behind this test and the outcomes in the next chapter.

3.3.3 Quantitative data collection for the Indonesian case study

A survey was designed to cross-check the results derived from the interviews, identify and remedy biases, and to obtain insight on issues emerging from the interviews that needed additional research, and that were suitable for a quantitative method. I used random stratification (Fink, 1995) as a sampling method and administered seventy-one guided questionnaires (Table 3.2). The presence of an interviewer during the guided questionnaires offers opportunities and risks. The most important opportunity is the fact that an interviewer can give direct support to the task performance of the respondent. For complex questions this is suitable. The most important risk is the influence or effect that the interviewer may have on the respondent's answers (Loosveldt, 2008). The key principle of standardised interviewing specifies that all questions are asked in the same way and that the respondents' interpretation of these questions is the same (Groves, 2004). This minimises the sample error. In this case the choice for guided questionnaire sessions was made to ensure quality and moreover to accommodate the need of respondents for some explanation of the questions, and to ensure that people gave serious consideration to the subject. This choice for quality has however limited the number of questionnaires that could be administered.

Instead of providing a qualitative answer as done during the semi-structured interviews, respondents were asked to rate the competences acquired during post-graduate education on a Likert scale, and the competences necessary to execute daily tasks, on a scale ranging from 1 to 5, where 1 = not at all, 3 = to some extent, and 5 = extensively.

Furthermore the respondents were requested to rate their use of various knowledge acquisition mechanisms along a similar scale and the follow-up of IPE and respond to a number of statements about organisational capacity. The survey was conducted in February and March of 2011 and can be found in Annex B.

3.3.4 Respondent selection in the Dutch case study

In the Dutch case study, 6 water sector experts from across the Dutch water sector were selected for the semi-structured interviews, bases on a consensus that they would be the best persons to provide a coherent and balanced overview of the sector. The sample population for the survey is the total group of water professionals with an MSc degree or higher. Approximately 80.000 Dutch professionals work fulltime in the water sector (2012); however this also includes professions with lower education than a MSc degree; so that the exact number of professionals with an MSc degree or higher is difficult to determine. In my sample I only include professionals with an MSc or higher from the following actors:

- Knowledge institutes (public and private research and educational institutes)
- Private organisations (contractors, consultancies, drinking water companies)
- The public sector (ministries, provinces, municipalities, water boards)
- Civil society organisations working on water management

I further choose to look at the following subsectors:

- Water for agriculture
- Integrated water resources management
- Engineering and construction
- Drinking water and wastewater
- Water quality and ecology
- Water quantity (hydrology, flood management)

I have chosen to sample the whole water sector and not only the Rijkswaterstaat because the Dutch water sector is more diverse than the Indonesian one, in terms of the number of professionals working for different actors. In Indonesia the DGWR is the largest player in the sector, whereas in the Netherlands the relative number of water professionals working for other actors than the government is much larger, and the diversity in the water sector may influence KCD as well.

To reach respondents in the sample population, I have relied on invitations on websites, newsletters, magazines and conferences, on invitations in online mailing lists and discussion groups that are used or read by the target group, and individual invitations (Table 3.3).

Table 3.3. Fora in which an invitation for the questionnaire was spread

Organisation name	Type of organisation/medium	No. Of members	Field
Waternetwerk	Professional Association	4000 members	Technical

(NVA/KVWN)			
Waterforum	Portal for water managers in industry, knowledge institutes, government	Newsletter to 13.000 professionals	Technical, management
Hydrology.nl (Unesco-IHP)	Dutch portal for hydrology and water resources, coordinated by the Dutch National committee for the Unesco-International Hydrological Programme.	n/a	Technical
Netherlands water partnership (NWP)	Networking organisation for the water sector	Newsletter to 3000 water professionals	Technical, management, governance
Dutch Hydrological Society (NHV)	Society for hydrology	600 members	Technical, management
The Rijkswaterstaat	Government	Individual invitation to 40 professionals	Technical, management, governance
Author	Personal network	Individual invitations to 30 water professionals	Technical, management, governance
LinkedIn groups: Unesco-IHE, KIVI-NIRIA, Water Professionals, Human Capital Water & Delta.	Social online media	Invitation to all group members	Technical, management, governance
Annual event for the water construction sector (Waterbouwdag 2010)	Conference	Invitation to audience	Technical, management

3.3.5 Qualitative data collection for the Dutch case study

The Dutch case represents a society where we hypothesise that the set of KCD mechanisms is more diverse than in Indonesia, and the case acts as an illustrative comparator case. Accordingly it required a smaller number of interviews. The arguments for the hypothesis are explained in the case description/Dutch case chapter. The interview questions were similar to the questions asked in the Indonesian case, but with minor alterations for the Dutch situation. They can be found in Annex C. Each interview lasted between 1 and 2 hours. The interviews were recorded, transcribed and subsequently anonymised.

3.3.6 Quantitative data collection for the Dutch case study

The interviews were followed by a quantitative survey. In contrast to the Indonesian case, the survey was conducted online. Online surveys are internet surveys that are computerised, self-administered questionnaires, answered without the presence of an interviewer (Manfreda and Vehovar, 2008).

I use an online survey because in the Netherlands people can be reached easily by internet and e-mail. Nearly everyone in Dutch society has access to a computer with internet. The disadvantages are the often low response rates of online surveys with general invitations (Manfreda and Vehovar, 2008). A possible bias may arise in that I had to rely partly on media that I want to investigate in the survey, meaning that I automatically reach the respondents that make active use of these KCD tools. Except for the choice for an online survey, the methodological approach and the type of questions were similar to the approach chosen in the Indonesian case, albeit the questions were adjusted to the wider Dutch water sector. The survey questions are presented in Annex D.

3.3.7 Qualitative data analysis for both cases

Following the completion of the interviews, Atlas TI version 6.2, a qualitative data analysis software package, was used for interview analysis for both the Indonesian and Dutch case study. Each interview was coded thematically according to the method proposed by Saldaña (2009) using codes based on the KCD conceptual model (Chapter 2). In addition, open coding was used to identify topics that emerged during the interviews and that could not be classified adequately using the thematic codes. This method choice means that a traceable route from qualitative data to its interpretation is established. By cross-referencing the thematic and open coding with the seven strata of the Indonesian case, a rich and composite understanding of the evolution of the four core competences within the DGWR in relation to the institutional context and post-graduate education was obtained. For the Dutch case a similar procedure was followed, based on the interviews with Dutch experts. Throughout the results chapters I use quotations from respondents to confirm or emphasise my findings. Each interviewee has a unique number, which ensures their anonymity, as agreed on during the interviews. This leads to relatively unbiased statements. The list of thematic and open codes can be found in Annex E.

3.3.8 Quantitative data analysis for both cases

The results of both surveys were analysed per strata, using the Statistical Package for the Social Sciences (SPSS) and STATA software (version SPSS Statistics 20 and STATA 12). To analyse the extent to which variation between groups of respondents can be attributed to real differences instead of random fluctuations in the sample, tests of significant difference and analyses of variance (ANOVA) were performed. Differences between cohorts or between local and international post-graduate education that are significant, are indicated with an asterix respectively, and reflect a statistical difference between these categories. Regarding the other items, it cannot be excluded statistically that differences are attributable to chance; however, they still indicate patterns.

4 An institutional analysis of the DGW

4.1 INTRODUCTION

In this chapter I focus on the institutions, the rules in use that define the actual behaviour in the DGWR and the influence they exert on KCD. In the adapted KCD conceptual model, this is the level of the enabling environment.

I do this by investigating how knowledge and capacities are used in the policy making process and how national and organisational culture determine knowledge exchange. Policy domains tend to stability because they are captured by groups of actors who share an interest in maintaining the status quo and who resist attempts to change prevailing policies and policy programmes. New knowledge becomes accepted in such a system to the extent that it fits the accepted discourses, as described by Litfin (1994). We investigate this process over a period of 40 years in the DGWR. Based on literature about the development of the water sector in Indonesia, transitions and policy change I distinguish moments where the dominant policy coalitions changed. Each coalition works under a different paradigm, requiring a different type of knowledge and capacity.

In Section 4.2 I will describe the complementary theory that I adopt to explain the relationships between the institutional settings and KCD. I explain the concept of institutions, the Multiple Streams Framework (MSF), Advocacy Coalition Framework (ACF) and the typology for national culture by Hofstede (1991). In Section 4.3 I analyse the results according to the theory introduced and Section 4.4 presents the conclusions of this chapter.

4.2 THEORETICAL EMBEDDING

4.2.1 Institutions

In North's seminal book (1990), institutions are defined as the rules of the game, the humanly devised constraints that shape political, economic and social interaction. He distinguishes both informal constraints (sanctions, taboos, customs, traditions, and codes of conduct), and formal rules (constitutions, laws, property rights). Throughout history, institutions have been devised by human beings to create order and reduce uncertainty in exchange (North, 1990). The institutional norms and arrangements largely determine what people talk about as well as who talks to whom in organisational processes (Schmidt and Radaelli, 2004). The way institutions are perceived has developed gradually and is continuously under discussion.

Ostrom (1992, 2005) views institutional development as a process that can be crafted consciously. Her Institutional Analysis and Development (IAD) framework has helped shape institutional analysis in various fields. It offers a set of criteria that are useful in institutional design. Bruns (2009) points out that designing institutions regardless of their relevance or feasibility in particular conditions should be avoided. In designing institutions, the fact that every situation is unique and needs its unique solution must be taken into account. Cleaver (2002) argues that people continuously and conveniently choose which institutions they use. This is not necessarily a conscious process. According to Cleaver and Franks (2005), 'Institutions elude design', making it problematic to assume design as conscious, rational crafting by narrowly self-interested actors (Bruns, 2009). Instead, Cleaver introduced the concept of institutional bricolage to suggest how mechanisms for resource management and collective action are borrowed or constructed from existing institutions, styles of thinking and sanctioned social relationships (Cleaver, 2002). While her argument reflects a close observation of reality, she does not provide an analytical framework for institutional development. Instead she deepens understanding of the means by which institutions can deviate from their intended ('designed') purpose.

4.2.2 Paradigm shifts

Kuhn (1962) described paradigm shifts in his seminal book on scientific revolutions, and compares them with political revolutions. Political revolutions are inaugurated by a growing sense, often restricted to a segment of the political community, that existing institutions have ceased to adequately meet the newly emerging problems posed by an environment that they have in part created. In the same way, scientific revolutions are inaugurated by a growing sense, again often restricted to a narrow subdivision of the scientific community, that an existing paradigm has ceased to function adequately in the exploration of an aspect of nature to which that paradigm itself had previously led the way. The sense of dysfunction that can lead to crisis is prerequisite to revolution. However, institutions are often geared towards preserving the status quo and thus towards optimisation and protecting investments, rather than system innovations (van der Brugge et al., 2005). Meijerink and Huitema (2009) describe this phenomenon in the policy making process, stating that policy domains tend to stability because they are captured by groups of actors who share an interest in maintaining the status quo and who resist attempts to change prevailing policies and policy programmes. New knowledge becomes accepted in such a system to the extent that it fits the accepted discourses, as described by Litfin (1994). The policy process is seen as the site of politics, processes of contest, negotiation, marginalisation, with knowledge production and use entwined with these forces: knowledge can serve to add legitimacy to political action often after the decision, and what counts as 'legitimate knowledge' is itself politically determined (Jones, 2009). A theoretical approach anchored in this assumption is the Multiple Streams Framework (MSF), by Kingdon (1995).

4.2.3 Multiple Streams Framework

Kingdon (1995) introduced the concept of 'windows of opportunity' to the policy sciences - a combination of favourable circumstances that creates an opening to influence the official policy agenda. He describes the policy process as follows: 'We conceive of three process streams flowing through the system – streams of problems, policies and politics. They are largely independent of one another, and each develops according to its own dynamics and rules. But at some critical junctures the three streams are joined, and the greatest policy changes grow out of that coupling of problems, policy proposals, and politics' (Kingdon, 1995).

The problem stream involves problem identification and recognition. These are based on focusing events or feedback from the system being managed or governed. Indicators can illustrate a problem's presence while focusing events can more broadly and immediately raise awareness of a problem.

The policy stream is seen by Kingdon as a 'primeval soup' (Bhat and Mollinga, 2009) in which policy alternatives are floated among and shortlisted by a community of specialists, based on their feasibility and acceptability. It requires knowledge of the formal rules, charting what already exists, formed opinions and associations (Jones, 2009); it is the official agenda (Meijerink and Huitema, 2009). Policies are proposals for change based on the accumulation of knowledge and development of interest among the specialists in a policy sector. Policy entrepreneurs propose solutions to policy problems (Gidron and Bar, 2009).

The political stream takes into account shifts in public opinion, changes in political administration, political parties trying to capture and maintain power.

What leads to coupling of the streams? The policy entrepreneurs; They are the advocates who are willing to invest their resources – time, energy, reputation, money, in return for anticipated future gain. These are people with a high ambition level, who have a claim to a hearing, have political connections or negotiating skill and who are persistent. The policy entrepreneurs must develop their ideas expertise and proposals well in advance of the time that the window of opportunity opens (Kingdon, 1995). During the pursuit of their personal purposes, entrepreneurs perform the function of coupling the previously separate streams. They hook solutions to problems, proposals to political momentum, and political events to policy problems (Kingdon, 1984).

4.2.4   The Advocacy Coalition Framework

The ACF is inspired by Kingdon's theoretical approach but particularly emphasizes learning over longer periods, in policy processes by a group rather than individual entrepreneurs. Advocacy coalitions, of individuals, organisations, or institutions, share a common set of beliefs about a policy problem and demonstrate some level of interaction (Sabatier, 2007). Knowledge plays a crucial role because the coalition is a reflection of the ideas and attitudes about a set of policy issues, the policy core beliefs.

Three theoretical lines of inquiry are important in the ACF, namely a focus on the nature of the advocacy coalitions, their structure and stability; the study of policy-oriented learning; and the role and behaviour of coalitions in policy change (Weible et al., 2011).

Further, policy change needs to be observed over a decade or more, to find out how policy analysis shapes the agenda and how learning takes place (Gidron and Bar, 2009). The ACF serves to guide theoretically driven inquiry about the extent to which people learn and what role scientific and technical information plays in policy making, and what factors influence both minor and major policy change (Weible et al., 2011). It has also proven useful for the analysis of water policy processes (van Overveld et al., 2010; Albright, 2011).

According to Sabatier (Albright, 2011), it is through alterations in factors external to the policy subsystem that policy core beliefs of competing advocacy coalitions may change. These external factors are divided into two broad categories: stable parameters and changing external system events. The stable parameters include the basic attributes of the policy problem; (ii) the distribution of political resources; (iii) fundamental social values and structure such as national culture; and (iv) the basic legal structure. Alternatively the events external to the system that are more prone to change over time include (ii) socio economic conditions; (ii) changes in governing coalitions (iii) policy changes in sectors external to the subsystem. This can also be described as the institutional environment (see Chapter 2).

I choose to adopt both the ACF and the multiple streams framework in my analysis as together they can provide insight in the water policy changes in the water sector of Indonesia and more specifically in the MPW, and the role of knowledge and learning in these processes, since the beginning of the 1970s.

In the next section, I introduce theory on national and organisational culture. This is imperative for deepening of understanding of the fundamental social values and structure necessary for a comprehensive explanation of KCD in the DGWR.

4.2.5 National and organisational culture

The society's national and organisational culture is important to understand institutional change. Culture is a conveniently inclusive term used to describe what a group of people share, including both tangibles and intangibles such as histories, traditions, symbols, ideas, values, attitudes, rules and achievements.

To avoid confusion I note that in this chapter I am interested in the individual level only in relation to the institutional environment, not in the individual level as such. The individual level will be investigated in the chapter on competence formation and education.

Several typologies of nation-level values and cultural and behavioural systems have been proposed and extensively studied. The seminal work by Hofstede (1991; 2001) has been followed by other projects; the Global Leadership and Organisational Behaviour Effectiveness (GLOBE) project in (House et al., 2001; House et al., 2004), the World Value Survey (Inglehart, 2004; Inglehart and Baker, 2000; Minkov, 2007), and the Schwartz cultural values model (Schwartz and Sagiv, 1995; Schwartz, 1992; Knafo et al., 2011). All these theoretical approaches provide important cornerstones in our understanding of cultural values. Despite the differences in the theoretical conceptualisations of nation-level value dimensions, and the variety of measures employed to measure them, the models of cultural values partly overlap, both conceptually and empirically (Knafo et al., 2011). Inglehart (2000; 2004) established the World Value Survey and has written on cultural values in relation to modernisation. Inglehart and Welzel (2005) distinguish culture along the lines of traditional vs. secular-rational values and survival vs. self-expression values, hypothesizing that modernisation will cause societies to move towards a more secular-rational worldview with more attention for self-expression.

While taking note of the criticism of Hofstede's approach (McSweeney, 2002a; Minkov and Hofstede, 2011; Jones, 2007; Baskerville, 2003; McSweeney, 2002b), the components of Hofstede offer a more operational tool for analysing of our empirical material than the dimensions of Inglehart and Welzel (2005) or these of Schwartz (1992). My aim is not to determine the best conceptualisation of culture (Enserink et al., 2007), but to find a cultural framework that will help in understanding the interplay between the institutional dynamics and the role of knowledge in the DGWR.

Following Schein (1985), one should bear in mind that only statements on specific elements or dimensions of culture can be made, and culture cannot be explained as an entity. Indonesia is a multi-ethnic country with around 900 permanently inhabited islands, spanning more than 5000 kilometres from east to west. Indices of culture that attempt to describe a 'national' common denominator, though workable, will always only be an approximation of reality. Organisational culture in turn is partly determined by the national culture, but is also determined by the type of work people do.

Hofstede (1991; 2001) identifies four dimensions that characterise culture:

1. Power distance, which is a measure of the degree of equality or inequality in a society;
2. Individualism/collectivism, which indicates whether individual or collective rights are prominent;
3. Masculinity, which is the degree to which society reinforces the traditional masculine work role model of male achievement, control, and power;
4. Uncertainty avoidance, which is the level of tolerance for uncertainty and ambiguity. If it is high, a country has a low tolerance for uncertainty and ambiguity and is rule oriented to reduce the amount of uncertainty.

Research on engineering culture shows that engineers have a strong inclination to eliminate uncertainty, by introducing technical, preferably people-free solutions, and tend to think in linear, simple cause and effect relations (Schein, 1996). In the case of uncertainty avoidance, therefore, the engineering subculture of the MPW may be dominant over the national culture.

I received secondary data from a resource person in the DGWR. The resource person issued personality tests based on the DISC quadrant behavioural model, that were filled in by 54 water professionals in the DGWR. The model is developed by Marston (1928), who described human behavioural styles in four categories, namely Dominance, Influence, Steadiness, and Compliance. He conceptualised people's behaviour as determined along two axes, with their attention being either passive or active, depending on the individual's perception of his or her environment as either favourable or antagonistic (Duck, 2006). By placing the axes at right angles, four quadrants are formed with each describing a behavioural pattern (Figure 4.1):

- Dominance produces activity in an antagonistic environment
- Influence produces activity in a favourable environment
- Steadiness/Stability produces passivity in a favourable environment

- Compliance produces passivity in an antagonistic environment.

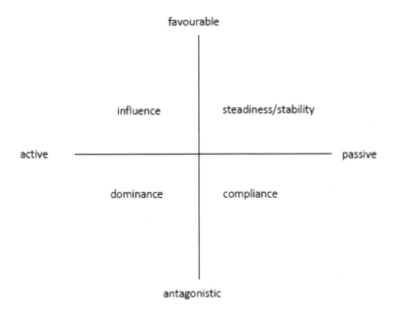

Figure 4.1. DISC quadrant behavioural model

Dominance looks at how an individual deals with problems. The higher the 'D' factor, the more aggressive and determined an individual will be in seeking solutions. High 'D' people are described as demanding, forceful, egocentric, strong willed, driving, determined, ambitious, aggressive, and pioneering. Low 'D' scores describe those who are conservative, low key, cooperative, calculating, undemanding, cautious, mild, agreeable, modest and peaceful. The 'I' factor, influence, addresses how individuals interact with people. People with high 'I' scores influence others through talking and activity and tend to be emotional. They are described as convincing, magnetic, political, enthusiastic, persuasive, warm, demonstrative, trusting, and optimistic. Those with low 'I' scores influence more by data and facts, and not with feelings. They are described as reflective, factual, calculating, sceptical, logical, suspicious, matter of fact, pessimistic, and critical. High 'S' individuals tend to be calm, relaxed, unemotional, patient, possessive, predictable, deliberate, stable and consistent. Sudden change is unwelcome. People with high 'C' styles adhere to rules, regulations, and structure. The higher 'C' factored people will be more rigid in their approach (Furlow, 2000).

In the Results and Discussion, I will utilise the four dimensions of Hofstede and the DISC model in analysing the Indonesian case.

4.3 RESULTS AND DISCUSSION

Now that I have created a cultural lens that enables us to understand how national and organisational culture moderate the uptake of knowledge and capacity in the MPW, I further discuss how knowledge and capacity played a role in policy change in Indonesian water management in the three phases. Information sources for this analysis include policy documents, consultant reports, legal documents, scientific literature and interviews.

In their book on policy entrepreneurs, Huitema and Meijerink (2009b) research the way policy science literature conceptualizes policy stability and change. A distinction is made between incremental change vs. radical fundamental deep or paradigmatic policy change (John, 1998, Huitema et al., 2006, Schlager, 2007 in Meijerink and Huitema, 2009). Most theories argue that under 'normal' circumstances policies are quite stable and only develop or change incrementally. North (1990) postulates that the kinds of knowledge, skills and learning that that members of an organisation will acquire will reflect the payoff - the incentives - imbedded in the institutional constraints. The knowledge acquired in turn shapes the institutions. This means an equilibrium in which an organisation such as the DGWR, providing incentives to continue construction oriented management, is receptive only to knowledge that fits the predominant paradigm. This can exist for a considerable time.

I hence identify three different paradigmatic phases in Indonesian water management, based on differences in political climate, new developments in water management and resulting differences in the focus of the water sector, leading in turn to notable changes in the MPW and DGWR. Consequently, knowledge and capacities are valued differently in each phase. For convenience, I term the water management phase before 1987 'Phase I', between 1987 and 1998 'Phase II', and after 1998 'Phase III'. The distinguishing features of these phases are explained in the next sections.

4.3.1 Phase I: 'Pembangunan' - Development of infrastructure: 1970s to 1987

Phase I is marked by the authoritarian and top-down rule of General Suharto, who was in power from 1966 to 1998. He exerted tight control on the civilian bureaucracy, employing the political party called Golkar as his political vehicle (Sarsito, 2006). Though Golkar was presented as a functional group, even the Indonesian press called it the government's party and regarded it as the political arm of the armed forces (Bird, 1998). For a civil servant to have a successful career, his or her loyalty needed to lie with the Golkar party.

Under Suharto, the generation of economic resources was of critical importance, with foreign-educated technocrats filling ministerial posts of economic ministries such as Bappenas, and the ministries of finance, trade and the MPW. They brought in technical and management knowledge from well-known international universities, partly because this knowledge was scarcely available locally.

It was attractive to work in the civil service because salary supplements from project implementation were an important part of the remuneration of most civil servants. There were supplements to compensate for inflation and honoraria for projects, which represented the largest single source of legally sanctioned supplementary income and were paid out of development (or discretionary) budgets. This indicates the strong incentive for civil servants to work in ministries that had large development budgets and in particular in those directorates that were in charge of project implementation. Similarly a strong incentive existed for line ministries to seek to increase the size of their development budgets (Booth, 2005).

The principles of water resources and irrigation sector management in this phase were based on Law No. 11 of 1974 on 'Water Resources Development' (UU 11/74). The state controlled the water resources as well as all development thereof, but there was room for delegation to province and district levels, without a shift of authority. Good governance in the form of public transparency and public participation were not specifically reflected in the law. Basic tasks and principles of water resources management were stipulated in Government Regulation (PP) No. 22 of 1982 'Water Management' (Herman, 2007).

The first phase is characterised by a strong preference for water resources development, that is to say a focus on the construction of infrastructure. This is understandable as it was imperative to build up transportation, communication and water infrastructure (flood protection works, hydropower, irrigation, water supply), for economic development and food security. The 1970s saw steady real growth in government revenues and expenditures, both in absolute terms and relative to Gross Domestic Product (GDP). Much of this growth was due to a rapid increase in government revenue from oil exports. At the peak of the second oil shock in 1980/81, total government expenditures were around 24% of GDP, the highest percentage in Indonesia's post-independence history. This figure declined through the 1980s as oil revenues declined, and the development budget contracted from around 12% of GDP (half of total budgetary expenditures) in the early 1980s to 8% by 1989. This illustrates the optimism about the development of the country in the beginning of the first phase. Because there were few engineers or professionals, the civil service had to work in a very structured manner and efficiently to attain the development goals. The lack of (technical) capacity at the local level was an important consideration in maintaining a centralized administration system. The structured approach was very productive: in 1985, rice self sufficiency was attained, leading to an award by the food and Agricultural Organisation (FAO). Interviewees exhibit a sense of pride when they refer to this period[4] and resource persons confirm these feelings[5].

The predominant advocacy coalition in the MPW was a construction oriented technocratic coalition including mostly professional staff, built on the belief that irrigation infrastructure development was important for the nation's development, and that it was best to keep control centralised and top-down. This was a very strong self-enforcing coalition. There was little participation from other parties, except from the international donor community. Because control was in the hands of a few highly placed, like-minded officials, decision-making was not complicated. Accountability was not required, because neither political opposition nor civil society had a strong voice, leaving room for large-scale corruption by officials who were part of the ruling party.

By the end of the first phase, the sector faced a declining physical and fiscal sustainability of existing river and irrigation infrastructural assets. Infrastructure had deteriorated

---

[4] PD44, PD46, PD57,
[5] PD32, PD33, PD48

because of pervasive under-funding of asset management, rectified by premature and expensive rehabilitation (Alaerts and Herman, 2005).

However, rehabilitation of existing irrigation schemes did not address weaknesses in maintenance capabilities and funding: some of the schemes rehabilitated in the early seventies were in urgent need of re-rehabilitation by the mid 1980s. In 1987, the first systematic effort was made to shift the emphasis from development to management of water resources systems—particularly in the irrigation sector, but also in river basin management (Houterman et al., 2003). The fiscal crisis caused by the collapse of oil prices in the mid-1980s highlighted problems with construction-oriented policies. The voice of those in government concerned with the need for increased attention for operation and maintenance (O&M) was strengthened. The WB and ADB sought to organise a local epistemic community, influencing relevant governmental officials through knowledge from the international discourse on participatory and integrated water resources management. The donors organised a policy forum (check Albright and Sabatier) by means of inter-ministerial dialogues, for the relevant government ministries to come to consensus on the nature of the problems. This resulted in the 1987 Irrigation Operation and Maintenance Policy (IOMP) (Herman, 2007), viewed as the culmination of the first phase. Irrigation, river basin management and water supply are the subsectors in which the users and stakeholders were supposed to acquire a more prominent role. However, under the authoritarian government of the time, this was not possible. The international donors can be typified as an advocacy coalition, although their advocacy eventually did not generate the results that they desired (Bhat and Mollinga, 2009).

4.3.2 Phase II: 1987 - 1998

This phase starts in 1987, the year the IOMP was introduced. Even though this policy did not really change construction oriented behaviour, it preluded a distinct next phase. By 1992, the MPW contemplated the introduction of integrated water resources management, some form of apex sector coordination institution, a national water resources policy, along with provincial river basin management units (*Balai PSDA*) (Alaerts and Herman, 2005). An international seminar, jointly sponsored by a number of international donors, on 'Water resources for Sustainable Use in Indonesia' held in 1992 at Cisarua in West Java, was important in recommending the application of IWRM principles to deal with sector problems. However, in practice the implementation of the dialogues and objectives of the IOMP, in the form of the Java Integrated Water Management Project (JIWMP) and Capacity Building Project (CBP) projects proved unsuccessful. Given the non-construction orientation of these projects, tasks for these projects were assigned to low-grade positions (Bhat and Mollinga, 2009). Irrigation, River Basin Management and Water Supply were fields in which the users needed to acquire a growing stake. However, this was not yet possible under the authoritarian regime.

Even though the IOMP was developed, indicative of a policy change, the change was only of an incremental nature and the incentives provided by the existing institutions were stronger, eventually leading to a lack of success for the IOMP (Alaerts and Herman, 2005; Herman, 2007; Bhat and Mollinga, 2009). Also, the knowledge on IWRM and IOMP was largely exogenously sourced, so there was little ownership. The requisite knowledge may have existed in-house as well but was only located with professional staff that had just returned from IPE. They would not yet have enjoyed a high enough status to be influential in the policy processes.

IWRM and IOMP concepts did not become the new norm. They require the surrender of power to more stakeholders and a shift from an infrastructure construction to a management paradigm. Knowledge on IWRM and on Operation and Maintenance (O&M) was therefore contested, and value-laden (Litfin, 1994). Many people in all layers in the hierarchy of the ministry had monetary interests in ongoing construction, meaning the transaction costs (North, 1990) of changing to a different water management approach were too high. Incentives to be part of the Golkar party were strong, as only then was a successful career ensured. Criticism could only be vented in private, and foreign consultants often functioned as a channel for raising questions and proposing alternatives. The difficulty in ventilating criticism forms an informal constraint and a disincentive to creating new knowledge and to use the knowledge that people possess. The fundamental pillars of Phase I stood straight until 1997. For a policy change to occur at the end of Phase II, the policy and problem stream would have to come together with the political stream. In this case the political climate hardened, with less attention for improvement of water management, decreasingly open to critical voices and becoming increasingly corrupt. . Therefore, policy change could not happen. However, Phase II can be regarded as a distinct phase because of the difference in political climate, making it progressively more difficult to do one's job.

The policy did not change the underlying dynamics of irrigation development because the government persisted in its role as operator, directly implementing activities, and trapping the system in old patterns. As the budget for rehabilitation remained considerable, this provided the opportunity to continue construction and rehabilitation and institutional development did not get the attention it needed (Bruns in Mollinga and Bolding, 2004). At the same time, irrigation projects continued to achieve formal targets as measured in terms of physical construction. International development agencies had strong incentives to continue to 'move money' and maintain good working relationships with the government, so their scope for pushing for major changes was limited, although their own criticism and dissatisfaction grew over time (Mollinga and Bolding, 2004).
Not much new knowledge on water management developed in this period. The time line of water legislation in Indonesia (Figure 4.2) shows the development of hardly any legislative developments in the second transition, hinting at a lack of innovation in the water sector over this period.
In Phase II, post-graduate education contributed by providing a new type of knowledge to a relatively small group of professional staff in the DGWR. The curricula of IPE programmes were changing at the time to pay more attention to the social issues related to water, and IWRM. However, professional staff of the MPW could not use their new knowledge in their positions, as it was contested knowledge that did not fit the accepted discourse. Meijering and Huitema (2009) point out that policy entrepreneurs require the skills to find appropriate venues to air ideas and to undermine the substance, procedures and organisations that work for the 'old' paradigm. These skills were not available in, nor taught in post-graduate education, at that time.

Government officials interviewed about this period expressed criticism when talking in private. 'The government officials usually felt unable to publicly voice ideas that might upset their superiors, though sometimes outsiders could function as a channel for raising criticisms and alternatives.' It first took changes in the leadership of DGWR and MPW in 1998 to open up the possibility of discussing institutional reforms in ways which could go beyond lip service (Mollinga and Bolding, 2004).

Phase II is also characterised by the growth of a civil society movement that could have resulted in a number of advocacy coalitions if it had had the chance to grow stronger. Better-educated young Indonesians worked in non-governmental organisations (NGOs) on issues like environment, the legal system, rural development, cooperatives, child labour, and so on. In effect, they undertook proxy activity for political parties. The movement was fragile, vulnerable to whims of government licensing and dependent on foreign largesse. It was never seen as a source of valuable knowledge or as an equal stakeholder in the policy-making process. It is only since January 2007 that civil society starts gaining a foothold in influencing policy-making processes through their role in the National Water Resources Council (Dewan Air), but their voice is not yet strong. The Civil Society Organisations (CSO's), such as the Institute for Social and Economic Research, Education and Information (LP3ES), often still have a status as subcontractors in projects, instead of as discussion partners or actors in policy processes. In Phase II NGOs represented a safety valve for a stale regime that increasingly lacked new or imaginative ways of continuing the process of lifting millions out of poverty and protecting Indonesia's extraordinary natural resources wealth (Bird, 1998). This is where the seeds were planted for the events of 1997 and 1998.

In the water sector, effective, integrated and sustainable water resources management, as envisaged by the Cisarua Seminar, had not been effectively implemented by 1998. As the country drifted into political and economic crisis, a number of institutions eroded further, as a result of the fading confidence in the leadership of Indonesia, speeded up by the agitation over seeing savings washed away, prices rising and jobs disappearing (Bird, 1998). People became more critical of the Golkar regime. Together with the financial crisis in Asia this eventually lead to the downfall of the regime, which marks the end of Phase II.

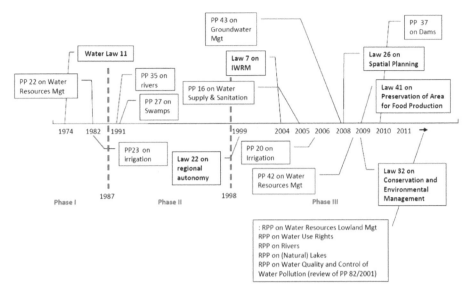

Figure 4.2. A timeline of relevant water resources management (WRM) legislation

4.3.3 Phase III: From 1998 until 2011[6]

The Suharto Government collapsed amid riots and high political uncertainty in May 1998. This opened a window of opportunity for significant policy change, as a political and problem stream came together (Kingdon, 1995). Large scale social change could occur because society as a whole was ready for a new paradigm (Kuhn, 1962).

In order to mitigate the crisis, the Government of President Habibie embarked on a strategy of policy and institutional reforms (Alaerts and Herman, 2005). The passing of laws on regional autonomy in April 1999 (22/1999 and 25/1999) laid the foundation for a major shift in power and money from the central government to districts. Decentralisation was ushered in as a way of fixing the unequal relations between the state and society and between centralised and local organisations, created by the highly centralized, patronage-based system of the Suharto authoritarian regime. Decentralisation was a political act, not an administrative one. The supporters of decentralisation needed to act faster than reactionary forces could regroup so as to retain control; administratively it was not well thought through. The decentralisation policy became effective on January 1st, 2001 (Usman, 2001). The radical turn to a decentralized administrative system may also have been driven by the need to ease mounting political pressures and to alleviate some of the administrative problems that contributed to the erosion of the government's legitimacy and control. As such, the decentralisation process was done without 'serious assessment and discussion across the regions and provinces as to what decentralisation would mean for them and the appropriate level of decentralisation' (Ramu, 2004). After decentralisation, problems naturally arose concerning management capacity and financial abilities at the provincial and district levels.

---

[6] Phase III was still ongoing at the end of data collection for this research

65

In the same period, BAPPENAS, the central planning agency, instituted a national forum for discussion of water resources sector reform to address the institutional lacunae in the water sector. The World Bank provided support through the Water Resources Sector Adjustment Loan (WATSAL). 'As a result of administrative and fiscal decentralisation, all sector management tasks and responsibilities were given to district governments except where a river or canal crosses a district boundary. The same implied if the river, lake, or canal crosses a provincial boundary, in which case, the national government may assume control' (Herman, 2007). The newly elected government also abolished the MPW in 1999 and replaced it with two ministries, namely the Ministry of Settlements and regional Development, and a State Ministry of Public Works.

In 2004, the government revised the water law after lengthy discussions, indicating a process of rethinking the decentralisation measures of 1999. The administration after 2004 resembles an administration system that is partly re-centralised (Schwartz, 2008) compared to what was envisaged in the laws 22/1999 and 25/1999. From 2005 onwards, 4 additional regulations have been written, on irrigation, drinking water, water resources management and groundwater. The law is designed according to the principles of IWRM, and contains many of the WATSAL reform agenda's objectives (Bhat and Mollinga, 2009). Irrigation Management Transfer (IMT) has been the most controversial of the reform components of the WATSAL because it would have curtailed the MPW's access to sectoral funds (Bhat and Mollinga, 2009), as the transfer of management would also mean a transfer of the maintenance funds.

DGWR had access to the global knowledge pool through links between the donor community and the MPW. This is also revealed by the influence the donor community tried to exert in implementing the WATSAL reform.

Bhat and Mollinga (2009) state that while the approach of building an 'epistemic community' after Haas (1992) proved unsuccessful for the purposes of implementing the WATSAL reform, it may still have an impact on the future of Indonesia's water resources sector, as the epistemic community includes pro-reform middle-management government officials, likely to occupy senior positions in future, replacing those with an interest in construction-oriented policies. This may not be the case if the group of officials is a minority in the MPW. Because of the zero-growth policy of the 1990s, the group of mid-level management officials is very small in general. However, the younger generation will have to grow into these positions fast and it tends to have a more critical mindset, and is more open to new ideas.

Law 7/2004 does not specifically address institutional development leaving the mandates for dealing with WRM scattered over different institutions. In addition to the MPW - DGWR, other organisations are involved, particularly the National Planning Board (BAPPENAS – for integration of planning), the Ministry of Agriculture (for food production and other agricultural water uses), the Ministry of Forestry (especially in the upper catchment areas), the Ministry of Environment (eco-system) and the Ministry of Energy and Mining (for ground water). At regional level similar agencies are active and involved in WRM. BAPPENAS and the regional planning boards coordinate the resource allocation for development activities in the different sectors, but are not involved in the coordination of the service implementation.

The MPW – DGWR has established Water Resources Management Units (*Balai or Balai Besar Wilayah Sungai ((B)BWS*) in national river basin territories (RBT), funded from the national budget. Funding from the provincial budget was used by the WRM agency to establish similar basin management units in the 1990s: the *Balai PSDA*. Coordination is the task of provincial water management councils (mandatory) and river basin councils (*TKPSDA*), but this is done only if deemed necessary by the stakeholders (Ministry of Public Works - Directorate General of Water Resources, 2010).

According to many respondents, competence (not only engineering competence), grew in importance in Phase III. However, one's political affiliation was still one of the determinants for higher management positions. This was mentioned by respondents[7] from Phase II/LPE, II/IPE, III/LPE, and resource persons: 'In this ministry, under this president, it is impossible not to choose your political colour'. These practices are hampering the development of the organisation, because positions are not necessarily occupied by employees with suitable expertise.

Bhat and Mollinga (2009) emphasise that the recruitment of staff outside of civil engineering to represent a wider set of skills in water resources management and restructuring incentives within the MPW to value these skills, build on the efforts made in the new policy streams. Until now, however, most of the recruited staff have a civil engineering background and the incentives to value other skills are only developing slowly.

As we have seen in this chapter the influence of knowledge brokers such as the donor community is limited, and they have only partly succeeded in establishing a community of reform-minded government officials, of which the majority are located outside the MPW. The influence of new knowledge is relatively small, because the new knowledge is contested by the vested powers. There is a clear hierarchy of knowledge, as is often the case in situations where technical expertise enables the monopoly by a single profession - in this case engineering - and is often reinforced by closed knowledge communities. Especially in a relatively authoritarian government such as Indonesia, the debate may be limited to the in-group, i.e. the advisors and aides to the person at the top (Weiss, 1999).

A number of features of policy making under the 'New Order' regime are likely to persist today, such as the strong hierarchy, the strong role for technocratic officials and little coordination with other organisations that have a role to play in water management, such as Bappenas. These features are reflected in the organisational structure, as discussed in the next chapter.

Suhardiman (2008) confirms this by stating that the characteristics of the Indonesian state remained largely the same in the aftermath of the regional autonomy. The state-citizen relationship continues to be shaped by the government's decision-making authority in development fund disbursement. Corruption practices continue to flourish, following the political reform in 1998. She also emphasizes that bureaucratic mechanisms of the present-day ministries continue to be shaped by patron-client relationships.

---

[7] PD37, PD51, PD53, PD57

4.3.4 National and organisational culture

4.3.4.1 Results of the DISC quadrant behavioural model

As described in the Methods Chapter, secondary test results were obtained of a DISC behavioural model test carried out in the DGWR in 2010 for 54 persons. The results indicate that 45% of the test takers has a tendency to the 'S' and 22% to the 'C' type with many respondents showing characteristics of the other type, e.g. test takers with a dominant 'S' often had 'C' as the second most dominant, and vice versa.

A high score for 'S' stability, means that people prefer to keep situations the way they were. Advocacy coalitions that promote other views and other knowledge that may lead to new policies will be perceived as uncomfortable.

The scores obtained with the DISC model will be dealt with in causation with Hofstede's characteristics in the following sections.

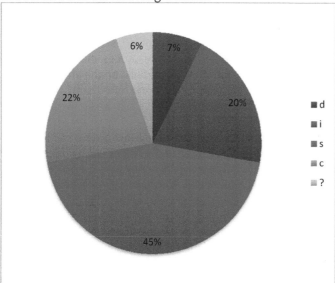

Figure 4.3. Results of the DISC personality model as tested in the DGWR in 2010

4.3.4.2 Power distance

Hofstede's work reveals that of the countries surveyed in his research, Indonesia shares the 8 and 9th place with Ecuador out of 50 countries in three regions, having a large power distance (Hofstede, 1991), compared with the 1st place for Malaysia, having the largest power distance of all researched countries, the USA on the 38th place and the last place for Austria.

A large power distance can be assumed to generally impede knowledge generation in multiple directions. It is argued that for both individuals to gain knowledge, a discourse, or argumentation with different opinions is necessary, such as a Hegelian dialectic, where a thesis and consequent antithesis lead to further developed knowledge, the synthesis, which is then in itself a new thesis (Kainz, 1996). This requires a minimum of openness to debate, and to appreciate differences of opinion. In the DISC model, this would translate mostly to higher values for 'D' and 'I' as they represent an eagerness to bring one's opinions into the limelight, and a desire for interaction. The 'D' factor is certainly not well represented in the DGWR and 'I' is overruled by a stronger tendency for modesty (low 'D') and diplomacy (high 'C').

The interview results[8] confirm the image of high power distance sketched by Hofstede. Some statements only emphasize the negative side of it: *'We are always afraid of the people above us. There is manipulation.'* And: *'The MPW is a very old organisation, based on seniority, rank, position, and sometimes you're not allowed to talk. I like to think and talk about my work but have heard senior officials say about me: If he would be my staff I would fire him.'* And, *'You cannot talk frankly and you cannot push. If you are not careful doing in how you approach your superior, it will have disastrous consequences.'* Remarkably, all such remarks were made by the respondents from Phase III, by an II/IPE respondent and several resource persons. The interview results[9] reveal a slight tendency for staff with an IPE background to find a modus for increased communication with seniors about the content of their work, that is that they apply the competences acquired during their IPE. The other respondents[10] tend to adopt a more passive attitude and more often seem to accept their work situation as it is, in line with a high 'C' score (Figure 4.3). In the interviews, IPE respondents[11] also indicated that they perceive their attitude as having changed while abroad, as explained in Chapter 6 on competence formation, whereas the LPE respondents state that their attitudes are ingrained and remain unchanged during their education.

Knowledge development literature indicates that much valuable learning happens informally on the job, in groups, or through conversations (Marsick and Watkins, 2001; Eraut, 2004). To support such learning, one needs to build a learning climate and culture. Climate and culture are built by leaders and other key people, who learn from their experience, influence the learning of others, and create an environment of expectations that shapes and supports desired results that in turn get measured and rewarded (Marsick and Watkins, 2003). Informal learning requires open communication, and is generally characterised by smaller power distances. In traditional Asian firms, and one may assume that this is valid in the public sector too, the 'superior' is expected to provide specialist expertise in technical matters (Rhodes et al., 2008); this high power-distance dimension deters subordinates from challenging the superior's opinion. In an environment of respect for power-distance and hierarchy, employees tend to follow the process of seniority, especially staff that are inclined to compliance (high 'C', Figure 4.3) Top-down decisions tend to go unchallenged. This suits a control and command leadership style and is effective to some extent, but may retard new ideas from subordinates, leaving little room for their empowerment (Rhodes et al., 2008). It furthermore impedes new knowledge to be easily used for policy processes. Without the supportive processes of involvement, ownership and empowerment, a high performance culture is difficult to achieve (Holbeche, 2007). The power distance may decrease in future, because the interviews show that the employees of Phase II and III tend to be

---

[8] PD27, PD32, PD33, PD35, PD36, PD38, PD39, PD42, PD50, PD56, PD58
[9] PD27, PD 51, PD62
[10] PD35, PD52, PD58, PD61
[11] PD27, PD30, PD38, PD40, PD43, PD51, PD62, PD64

more direct, open and critical than the interviewees of Phase I. The respondents of Phase I were raised and worked for a considerable amount of time under the Suharto rule, and may be more cautious in ventilating their opinions. Also, the older generation is very polite, and being critical is not common in Indonesian culture, this is the generation that is now positioned in leadership functions. Another more pragmatic reason is that for senior staff, often in higher management positions, there is more at stake than for young employees. The respondents of Phase III never worked under the Suharto regime and even though the organisational culture in the ministry has only slowly become more open, they demonstrate a freer attitude. Furthermore, their generation is an avid user of social media and internet. Much more information has become available since 1998 and with the use of internet the availability of information has grown exponentially. It has become much easier to check the validity of people's statements by comparing them with other sources. This is confirmed for the Indonesian situation by Tjakraatmadja (2011). Furthermore, when economic welfare increases, power distance usually decreases, because people become more independent financially.

However, until now, the strong orientation to maintaining power differentials and hierarchical structures (a high 'C', Figure 4.3), often has confined proprietary knowledge to elite cliques and inhibited organisational learning. Since a key governance tool is information secrecy and manipulation (Westwood, 1997) there is a strong tendency to keep knowledge non-codified and undiffused (Snell and Hong, 2011). This has considerable influence on policy change and enforces the status quo. Without felt pressure to codify and diffuse knowledge making it accessible outside immediate social networks, knowledge is likely to remain sticky, thus hindering implementation of the externalisation and combination processes (Nonaka and Takeuchi, 1995). Furthermore, deference to leaders may, for fear of reprisal continue to reduce subordinates' courage to challenge 'dominant logic' by offering ideas for change or improvement (Prahalad and Bettis, 1986), thereby inhibiting double-loop learning within the DGWR (Argyris and Schön, 1978).

4.3.4.3 Individualism vs. collectivism

Indonesia is ranked 47th/48th together with Pakistan on the individualism index. 53 countries or regions are ranked. The USA is ranked number one, being the most individualistic, compared to the group of Arab countries on the 26th and 27th place and a few South- and Mid-American countries which also score higher than Indonesia. Collectivist cultures entail a strong predilection to act as members of a group (Snell and Hong, 2011). In such an environment, the provisions of the organisation for the employee - training, physical conditions in the office, skills used on the job - are appreciated more highly than in an individualist environment. In this way the employee's dependence on the organisation is stressed. These things may be taken for granted in individualist countries, and therefore valued less. Indeed, the DGWR does not take an individual approach in the provision of Human Resource Development (HRD), for example. As will be elaborately discussed in the next chapter, the Human Resource Management (HRM) system is relatively mechanistic, and a personal staff development and appraisal system does not exist. Facilities are arranged for groups and individual career development is not arranged for. The staff tend to accept the working environment as it is, although respondents from all phases as well as resource persons criticise the lack of performance evaluation when talking in private, and staff employed in Phase III criticise the lack of connection between training needs and the training that is available. Frequently mentioned in this regard, specifically by IPE respondents from each phase returning from their education abroad, is that they have to find a way to adjust to the organisational environment again. They mention they have to choose whether they want to work in such an environment, and whether they want to be loyal again, whether they want to be careful in ventilating their opinion or knowledge. This manner of working inhibits knowledge exchange. Collectivism in the workplace is characterised by employees acting according to the interest of the group, even if this does not coincide with the individual interest. This is confirmed by the high 'C' value for compliance (Figure 4.3). The relationship between the employer and employee is seen in moral terms, it resembles a family relationship with mutual obligations of protection in exchange for loyalty. Loyalty to the group in a collectivist culture is paramount, and over-rides most other societal rules and regulations. Poor performance of an employee in this relationship is no reason for dismissal (Hofstede, 1991).

The society fosters strong relationships where everyone takes responsibility for fellow members of their group. This was clearly seen in Phase I and II, where in case one was loyal to the Golkar party, loyal to the organisation and to the superior, he or she would enjoy benefits. This may also lead to policy coalitions in the DGWR that are even stronger because of loyalty to the group and could further resist policy change that is pushed by other coalitions.

Resource persons indicate that they see DGWR and the ministry as a strong group, and employees that they feel part of a group, a family. This is confirmed by 4 resource persons, LPE as well as IPE respondents in each phase.

Performance evaluation in the DGWR functions in accordance with these principles. Average to high ratings are standard in the majority of employee appraisals; avoiding awarding poor employee ratings nevertheless translates into a lost opportunity for counselling underperforming employees. This undermines the performance appraisal system (Rhodes et al., 2008) and results in lost learning opportunities.

Hofstede (1991) suggest that when working in a collectivist surrounding where the group of colleagues functions as an emotional community, incentives and bonuses should be given to the group as a whole, not to individuals. This maybe an important recommendation for the bureaucratic reforms currently taking place in Indonesia.

4.3.4.4 *Masculinity vs. femininity*

Indonesia shares the 30 and 31st place with West Africa as a region where culture is perceived as a 'feminine' culture, together with the Netherlands, and in contrast with the USA and the UK. However, the type of work practiced in the MPW, focuses on individual technical performance, which calls for masculine values (Hofstede, 1991).

A very prominent feature of 'feminine' culture in Indonesia that can however be recognized in the MPW, is the attention for people and the importance of establishing relationships between people. Across much of Asia, because legal systems and institutional rationality differ from the West, intricacies of social relationships permeate all aspects of life, such that even in formal working contexts the quality and effectiveness of interpersonal cooperation depend on the strength of the social and kinship ties that bind the parties, along with the associated obligations. Such ties are inescapable considerations when interfacing with government bureaucracies (Snell and Hong, 2011). People need their relationships and their network to obtain information, as the formal system of information provision is imperfect. This was indicated as important in every phase, but in Phase II maintaining good relationships was increasingly important to demonstrate your loyalty to the regime, and also to make sure you were noticed in the distribution of informal benefits. For example,

...' studying abroad was more about the connections you had...' (resource person)
'...Political connections are important in Indonesia, if you want to get somewhere...' (Phase II respondent.

Pervasive personal networks can have substantial downsides at the societal level. The economic crisis in 1997 has been attributed to 'crony capitalism', a system in which successful businesspeople derive their success from close relationships with politicians and government officials (Snell and Hong, 2011).

In 'feminine' countries, both boys and girls learn to moderate their ambition and be modest. Assertive behaviour and attempts to excel are appreciated in 'masculine' cultures but are easily ridiculed in 'feminine' ones. In Indonesia, the feminine culture is shown in the need to blend in. Showing that you are knowledgeable on a topic needs a careful approach, because it is considered inappropriate to stand out. This concords with the behavioural style of a high 'C' factor, (Figure 4.3) leaving little opportunities for knowledge exchange.
Literature on the policy making process indicates that meetings are often held with a very large number of representatives, leaving little space for real debate and discussion. An unwillingness to be associated with particular positions, at least formally or openly, means that participants tend to avoid criticism of policy initiatives, and limits the extent of innovation and change (Datta et al., 2011).

4.3.4.5 Uncertainty avoidance

Indonesia scores relatively low (41/42nd place) on uncertainty avoidance, in Hofstede's work, compared with Germany (29) and Greece (1st place) for example. Weak uncertainty avoidance is characterised by accepting uncertainty as a normal feature of life; each day is accepted as it comes. Stress, aggression and emotion are not shown, and people are comfortable with ambiguous situations. Time is a framework for orientation (Hofstede, 1991), instead of a hard sequence of deadlines that governs actions by the minute. The interviews suggest an ambiguous picture for the DGWR. On the one hand they suggest that the ministry is characterised by a strong uncertainty avoidance, which is characterised by structured learning situations, giving the right answers, the need, apparently, for adherence to clear rules and regulations, suppression of deviant ideas and behaviour, and the need to blend in (Hofstede, 1991): the 'C' type in the DISC behavioural model. Respondents[12] expressed the wish to have a long term career in the MPW, indicating uncertainty avoidance. Also, several respondents[13] (resource persons, Phase III/LPE respondents) mention that MPW employees are generally not interested in involving different interest groups. The involvement of stakeholders, introducing an uncertainty factor, could 'spoil' a perfect linear, technical solution by introducing uncertainty. This is expressed by resource persons, by one LPE respondent from Phase II and Phase III. Similarly, creating more lateral communication with other organisations, allowing new knowledge to enter the ministry, has only increased since the onset of Phase III. At the same time, the weak uncertainty avoidance is recognised by allowing for ambiguous situations such as unclear distribution of tasks leading to neglected O&M. This has lead to dangerous situations, or disasters such as the Situ Gintung dam collapse.

4.4   CONCLUSIONS

In this Chapter, I have sketched the institutional environment for policy change and its relation with KCD in the DGWR, from the 1970s until the present.
I use policy theory to distinguish three different water management phases, showing how the institutional context, external pressures such as the political situation and national and organisational culture constrain the development, uptake and use of knowledge. I also address how advocacy by international donors and new insights, together lead to shifts in the approach to water management.

4.4.1   Knowledge for policy making

In Phase I, the predominant advocacy coalition in the MPW was a construction oriented technocratic coalition including most professional staff, built on the belief that irrigation infrastructure development was important for the nation's development, and that it was best to keep control at the central level. The technical expertise and available skills mix were appropriate for most of Phase I, although towards the end of Phase I, a need for a different kind of competence arose which was not allowed to grow. A system for critical review and corroboration of knowledge was not available. There was little participation from other parties, except from the international donor community. Knowledge brought in by the donor community was threatening to the status quo in the DGWR and the donor community did not manage to change the existing technocratic water policies of Phase I. The political context, although gradually becoming more closed and authoritarian did not really change, and thus did not provide a window of opportunity for significant policy change; the change from Phase I to II was only incremental, and not a paradigm shift in terms of policy. However, politically, there was a significant shift in the sense that it became increasingly important to be loyal to the regime at the cost of all aggregate competences and the meta-competence for continuous learning. Corruption increasingly flourished. At that time knowledge on IWRM may have started to come in through professional staff just returned from IPE. These alumni would not yet have a high enough status to be influential in these policy processes. IWRM and IOMP did not become the new norm in Phase II, as they require the surrender of power to more stakeholders and a shift from a building to a management paradigm, for which in Phase II

---

[12] PD27, PD35, PD37, PD38, PD58
[13] PD32, PD33, PD41, PD48, PD49, PD50, PD56

the competence was not available. Knowledge on IWRM and on O&M was contested, and value-laden. Knowledge only plays a significant role in the policy processes, if it fits with paradigm of that moment.

The shift from Phase II to III provided the window of opportunity to a new paradigm, in line with Kingdon's multiple streams theory. The fall of the regime, a major political event, opened the way to a decentralisation policy, under influence of strong public pressure to do so. This had lead to a gradual increase in governance competence.

A steady stream of internationally available knowledge has come in over the years through links with the donor community, international consultants, and IPE. However, the influence of knowledge coming in through these channels is limited. The channels may have a comparatively large role because the organisational system is relatively closed to other KCD mechanisms, but at the same time the donor community has only partly succeeded in establishing a community of reform minded government officials, of which the majority is located outside the MPW. Furthermore there is still a clear hierarchy of knowledge, as is often the case in situations where technical expertise entails the monopoly of a single profession - in this case engineering - and this is reinforced in the closed community of the DGWR. Technical knowledge as a basis is very much appreciated, supplemented by knowledge to run the administrative processes. This is elaborately described in Chapter 6 on Competence Formation.

The new paradigm of Phase III has resulted in a slowly increasing appreciation and acknowledgement of other types of knowledge, as is also seen in Chapter 6 on competences and post-graduate education. The influence of civil society is also growing, although slowly, opening-up avenues to incorporate knowledge from a variety of sources. This will put pressure on the government to be accountable, and will lead to increased innovation, represented by an increase in the meta-competence for continuous learning and innovation (Figure 6.2 and 6.3).

4.4.2   National and organisational culture

Hofstede's framework showed that Indonesia has a large power distance. High power distances generally impede knowledge creation because employees tend to follow the process of seniority and do not question top-down decisions. It is expected that the power distance may decrease because the employees of Phase II and III tend to be more direct, open and critical than the interviewees of Phase I.

The collectivistic characteristics of the DGWR show in the approach to HRM. Training is generally organised for groups, and personalised staff development programmes do not exist.

A high score for femininity is reflected in the preference for modesty, and the need to blend in, instead of standing out. For knowledge exchange this can be problematic because it is difficult to show that you are knowledgeable on a topic. Both the large power distance and the need to blend in inhibit the institutional capacity for learning, represented by the meta-competence for continuous learning and innovation. At the organisational and individual level this may be compensated by KCD mechanisms that fit the cultural traits.

5 Readiness for future challenges: organic vs. mechanistic organisational structure in the DGWR

5.1   INTRODUCTION[14]

In new democracies there is often a disconnect between the new democratic leadership and the administration it inherits, as the latter tends to live its own life guided by the old paradigms (Synnerstrom, 2007). This is certainly the case in Indonesia's young democracy. The Directorate General of Water Resources (DGWR) in the Ministry of Public Works (MPW) in Indonesia is an old, venerable organisation, with formalised procedures, rules, regulations and formalised communication throughout the organisation. In this chapter I focus on the role of the formal organisational structure in knowledge and capacity development (KCD). The organisational structure does not dictate how organisation members actually behave; it only provides formal guidelines and a framework (Christensen, 2007).

People generally behave differently than how an organisational structure prescribes. This has been researched as early as in the 1950s and 1960s (Dalton, 1959; Downs, 1967). They consciously or subconsciously make use of formal or informal rules and procedures to reach their goal. In the previous chapter I discussed the manner in which actual work is done in relation to policy making, with the aid of with institutional theory. In this chapter I want to discuss the relation between formal organisational structure and KCD, and the actual behaviour in relation to the formal structure, with the help of organisation theory.

Organisational structure is defined by Child (1972) as the formal allocation of work roles and the administrative mechanism to control and integrate work activities including those which cross formal organisational boundaries. The organisational structure sets limits as to who can participate in processes and encompasses role expectations and rules for who should do what, and how. The organisational structure further includes the formal rules governing these arrangements. Mintzberg has conceptualised five different organisational configurations, of which the DGWR is found to most resemble the Machine Bureaucracy configuration. The Machine Bureaucracy is related to the mechanistic management system introduced by Burns and Stalker (1961). In this chapter I investigate the extent to which the DGWR resembles the Machine Bureaucracy and the extent to which it has adopted a more organic structure appropriate to the present challenges of integrated water resources management (IWRM) and climate change. In addition, I investigate how staff in the DGWR work with the organisational structure. In the following sections I briefly describe theory on organisational structure and follow this with an explanation of the study method. The results of applying this to the DGWR are described and discussed. I will describe the formal organisational structure and formal rules, and also the actual behaviour in relation to the formal structure. In the last section, I draw a number of conclusions and provide recommendations on how to improve the alignment of the organisation to the present day challenges.

---

[14] This chapter is a revised version of a conference article published by Kaspersma et al. Kaspersma, J. M., Alaerts, G. J., and Slinger, J. H.: Readiness for future challenges: Organic vs. mechanic organisational structure at the DGWR in Indonesia, World Congress on Water, Climate and Energy - Building a sustainable global future, Dublin, Ireland, 2012b, 13, .

5.2 THEORETICAL FRAMEWORK

Mintzberg (1979, 1983) developed five basic organisational configurations that provide a framework to understand and design organisational structures. He describes the Machine Bureaucracy as: '... Highly specialised, routine operating tasks, formalised procedures, a proliferation of rules, regulations, and formalised communication throughout the organisation, large-size units at the operating level, reliance on the functional basis for grouping tasks, relatively centralised power for decision making, and an elaborate administrative structure with a sharp distinction between line and staff.'

Burns and Stalker (1961) have described the characteristics of mechanistic and organic systems (Table 5.1.), in which the Machine Bureaucracy is positioned at the mechanistic end. A mechanistic structure is appropriate to stable conditions whereas an organic structure is viewed as more appropriate for changing conditions. Changing conditions continually give rise to fresh problems and unforeseen requirements for action, which

Table 5.1. Characteristics of a mechanistic and an organic management system

	Mechanistic management system	Organic management system
Appropriate condition	Stable	Changing
Distribution of tasks	The specialised differentiation of functional tasks into which the problems and tasks facing the concern as a whole are broken down;	The contributive nature of special knowledge and experience to the common task of the concern;
Nature of individual task	The abstract nature of each individual task, which is pursued with techniques and purposes more or less distinct from those of the concern as a whole; i.e. the functionaries tend to pursue the technical improvement of means, rather than the accomplishment of the ends of the concern;	The 'realistic' nature of the individual task, which is seen as set by the total situation of the concern;
Who (re)defines tasks	The reconciliation for each level in the hierarchy, of these distinct performances by the immediate superiors, who are also, in turn, responsible for seeing that each is relevant in this own special part of the main task;	The adjustment and continuous redefinition of individual tasks through interaction with others;
Task scope	The precise definition of rights and obligations and technical methods attached to each functional role;	The shedding of 'responsibility' as a limited field of rights, obligations and methods (problems may not be posted upwards, downwards or sideways as being someone else's responsibility);
How is task conformance ensured	The translation of rights and obligations and methods into the responsibilities of a functional position;	The spread of commitment to the concern beyond any technical definition;
Structure of control, authority and communication	Hierarchic structure of control, authority, and communication;	A network structure of control, authority, and communication. The sanctions which apply to the individual's conduct in his working role derive more from presumed community of interest with the rest of the working organisation in the survival and growth of the

		firm, and less from a contractual relationship between himself and a non-personal corporation, represented for him by an immediate superior;
Locating of knowledge	A reinforcement of the hierarchic structure by the location of knowledge of actualities exclusively at the top of the hierarchy, where the final reconciliation of distinct tasks and assessment of relevance is made;	Omniscience no longer imputed to the head of the concern; knowledge about the technical or commercial nature of the here and now task may be located anywhere in the network; this location becoming the ad hoc centre of control authority and communication;
Communication between members of organisation	A tendency for interaction between members of the concern to be vertical, i.e. between superior and subordinate;	A lateral rather than a vertical direction of communication through the organisation, communication between people of different rank, also, resembling consultation rather than command;
Governance for operations and working behaviour	A tendency for operations and working behaviour to be governed by the instructions and decisions issued by superiors;	A content of communication which consists of information and advice rather than instructions and decisions;
Values	Insistence on loyalty to the concern and obedience to superiors as a condition of membership;	Commitment to the concern's task and to the 'technological ethos' of material progress and expansion is more highly valued than loyalty and obedience;
Prestige	A greater importance and prestige attaching to internal (local) than to general (cosmopolitan) knowledge, experience, and skill.	Importance and prestige attach to affiliations and expertise valid in the industrial and technical and commercial environment external to the organisation.

cannot be broken down or distributed automatically according to the functional roles defined within a hierarchical structure. In mechanistic systems, individuals will have specialised knowledge to work on a subcomponent of the overall tasks of the organisation. The job is abstracted from reality, as each individual in this system is working on a small part only. A mechanistic structure requires less communication because roles and responsibilities, and the extent of dependence on others are clearly delimited and provide a clear frame for action. Consequently, little knowledge is generated through work discussions. The focus is on the knowledge and expertise inside the organisation, whereas in organic systems there is a need to obtain external knowledge to be able to adapt to a dynamic environment.

In organic management systems, in proportion to the rate of change of the environment, hierarchy and central command are less appropriate (Burns and Stalker, 1961; Rainey, 2003). The job can only be finished by participating with others in the solution of problems. This places a heavier demand on the individual and requires intensive communication and extensive discussion with peers. As the environment becomes more fragmented, and organisation must reflect this complexity in its own structure, giving the people in the units that confront these multiplying environmental segments the authority they need to respond to the conditions they encounter (Rainey, 2003).

In more recent work than that of Burns and Stalker and Mintzberg on organisation science, theorists have sought to understand organisations through systems theory and organisational economics (Barney and Ouchi, 1986), by power and politics in organisation theory and institutions theory (Ostrom, 2005; North, 1990), by studying organisational culture and learning (Argyris, 1993; Schein, 1985; Senge, 1990), and from a social learning and adaptive management perspective (Chiva-Gómez, 2003; McDaniel Jr., 2007; Pahl-Wostl, 2002).

I adopt Mintzberg's and Burns and Stalker's foundational work because it appropriately describes the formal configuration of the DGWR. I argue that the organisation is structured in a mechanistic manner, and will need to move to a more organic structure, to be able to deal with current and future water management challenges. In the next sections I structure the results and discussion around the elements of organisational structure that are most important for KCD: the administrative divisions, the career system, and training and other KCD mechanisms.

5.3   RESULTS & DISCUSSION

5.3.1   Administrative divisions

The DGWR is marked by a high degree of formalisation and the centralisation of responsibilities, which is expressed in a strong hierarchy, division of labour and routines. Hierarchy entails superior and subordinate positions and various vertical levels in the organisation (Christensen, 2007). In the DGWR, a number of sections are answering to a sub-directorate, which is accountable to the Directorate, which in turn is responsible to the Directorate General (DG). Most of the power and authority concentrates at the Director and DG level, which is characteristic for a mechanistic structure. The DG level is where knowledge is accumulated and where the assessment of the relevance of distinct tasks is made (Burns and Stalker, 1961). Communication is vertical, in the form of instructions and assignments, from the level of Directorate General, to Directorate and Sub-Directorate or lower (Figure 5.1) and reporting back.

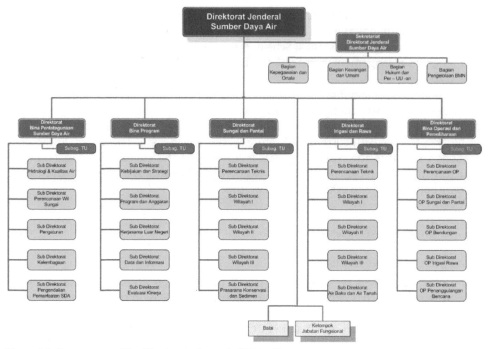

Figure 5.1. Organogram of the Directorate General of Water Resources

5.3.2 Career system

In Indonesia, civil servants typically enter the government at a young age and are guaranteed tenure, salary, promotion and other benefits. Promotion through the ranks is based on seniority and completion of training. Civil servants are allocated to positions through management decisions, not through competition among applicants (Synnerstrom, 2007), as in many countries (Wade and Chambers, 1980). The system does not allow lateral recruitment of mid-career professionals for limited terms, an important constraint to obtaining knowledge from outside the organisation (Datta et al., 2011).

The civil service in Indonesia also employs a comparatively large number of non-civil servant staff, in addition to the civil servant staff. The civil servant staff have permanent contracts and are part of the formal civil service system, the other staff fall outside this system and are often employed temporarily. The civil service staff are divided into line management positions and functional positions that are linked to specific technical professions. Civil servants are part of a career system, described elaborately by Tjiptoherijanto (2008), which divides staff into four ranks, from I (the lowest) to IV (the highest), each with a salary scale. Civil servants' ranks are based on their educational qualifications and seniority in the organisation. A direct relation of rank and grade with the echelon level for line management positions exists, i.e. the higher the echelon, the higher the rank and grade of the civil servant (Rohdewohld, 1995). Line management positions are considered more attractive than functional positions, as they provide better prospects for career advancement, professional status, financial and other benefits (Synnerstrom, 2007).

Since 2007, non-civil servant staff have the opportunity to apply for a functional position and become civil servants, and functional staff can apply for line management positions (1 resource person). The change to a new structure in 2007 is designed to facilitate faster promotion of staff. This reflects a move away from a purely mechanistic system towards a more organic structure, with enhanced flexibility in placing staff in appropriate positions.

5.3.3 Recruitment

In Phase I, recruitment was relatively straightforward. DGWR needed technical expertise and could hire a specified number of new civil servants each year.

In Phase II, and a few years into Phase III because of the financial crisis, Human Resource Management (HRM) in the DGWR was characterised by a zero-growth policy. The policy was introduced because the civil service was considered very large and inefficient. As a result, the average age of staff in the DGWR is high. Sixty percent of the staff are over 50 years of age, and within 3 years 43% of the DGWR staff will retire (Ministry of Public Works - Directorate General of Water Resources, 2010). Although DGWR has received new recruits in the past years, the number of recruits is not enough to compensate for the knowledge leaking away with the retiring staff in the opinion of resource persons, because it will take years before those new recruits will reach the same competence levels and productivity as the experienced retiring staff. This may continue possibly up to 2020. Thus, the DGWR needs to make detailed plans to retain the experience of the senior officials and replenish the quantity of workers to continue executing the tasks of the ministry.

Moreover, the DGWR has difficulty retaining new young staff. Of the 100 new recruits, on average 15 leave within the first year of service. Considering that being a civil servant is considered as a desired job, this is a very high number according to resource persons. An important reason for the early departure is the HRM system that is perceived as mechanistic leaving little space for the recruit's own ambitions and desires, as confirmed by respondents from Phase I, III and resource persons. For the respondents who were recruited in Phase III, this sentiment is more difficult to overcome than for employees who are now senior, as the younger generation has a more individualistic and merit-oriented attitude. Another reason for early departure is the low level of formal guidance on-the-job to get acquainted with the working environment. The MPW (2010) indicates that recruitment is a heavily centralised process (at MPW level) and that unit managers should become directly involved in the recruitment process. The unit managers need to prepare job descriptions, and do the final selection in cooperation with Human Resource (HR) staff. The HR staff are required to establish better relationships with unit managers and staff to be up to date about the knowledge and capacity needs in the units. Such an approach resembles a more organic management system, with a more personalised approach requiring horizontal communication across unit boundaries.

5.3.4 Financial & career incentives

The current HRM system has a formal performance evaluation system, a standard procedure that, however, tends to be regarded as lip service, according to several respondents, notably from Phase III, as well as resource persons. Several respondents from all 3 phases and resource persons also note that job performance is not explicitly taken into account in determining the salary. This limits incentives for employees to be creative and propose improvements (Rhodes et al., 2008). There are few tools to discourage low performance. In practice, actual performance evaluation is dependent on the preference of the supervisor. It is deemed satisfactory if one is an effective problem solver, and at the same time conforms to the informal and formal rules of the MPW, or as a respondent formulated it:

'You do not need to be clever in the sense of being clever in school; you need a different type of cleverness, to know how to work in such a system. MPW is quite a unique place, there is a system, a way of working. You cannot talk frankly and you cannot push'.

The actual rule is that if a staff member performs well according to the standard of the superior, he or she may be transferred to a position that includes involvement in many projects. Projects involve honoraria that top up the formal salary, and positions that include involvement in projects are popular. It can be concluded that strong incentives exist outside the formal basic ranking and salary scales that encourage performance according to the internal rules.

Several respondents argued that poor performance, absenteeism and corrupt practices can be justified by claiming underpayment:

...'Most of the time we are just sitting in our rooms, because the salary is very small'...

However, the view that performance will increase if the basic pay check would be higher is under debate (Mcleod, 2005; Filmer and Lindauer, 2001; Forest, 2008; Bowman, 2010). The Ministry of Administrative Reform in Indonesia, responsible for the bureaucracy reform, holds the view that a pay reform should be linked to better personnel management, while arguing that pay increases as such will not lessen corruption in the civil service (World Bank, 2005). Mechanisms for KCD

5.3.4.1 Education and Training

The organisation of training is taken care of by the Education and Training Agency of the MPW, by the Bureau for Personnel and Management of the Organisation under the Secretary General's (SG) office and by the Section for Personnel and Management of the Organisation under the Secretariat of the DG. Many respondents, mostly from Phase III, mention that training is sometimes ill connected to their needs, which may be attributable to the lack of communication between the units responsible for training, or to the fact that the training is offered to larger groups of people. However, the current high staff mobility at the DGWR makes the profusion of training necessary as well, for new staff to be easily introduced to new skills.

In each phase, the DGWR has sent a large number of staff abroad for IPE. IPE is important, as it is one of the few systematic and institutionalised mechanisms for KCD in the organisation, in addition to training. It is therefore unfortunate that a formal structure to follow-up on IPE and training of staff does not exist according to many interview respondents and resource persons[15], nor is it documented. The survey results show that in Phase II, follow up of IPE was rated much lower than in the other two phases (Figure 5.2). This could be because in Phase I, education was mostly technically oriented and technical knowledge only was appreciated in the MPW, resulting in relatively straightforward implementation. As shown earlier, knowledge was less important in government in Phase II, in contrast to loyalty to the prevailing political establishment. Less attention for follow-up of education is a logical result. In Phase III, knowledge is valued more highly again. For both Phases I and III, nearly all of the aspects are rated only slightly higher than three, which is a modest performance. For education and training support, many respondents indicate that follow-up depends largely on the personal appreciation of learning of one's superior. The dependency on the superior for any follow-up shows the relatively low priority in the organisation for learning and

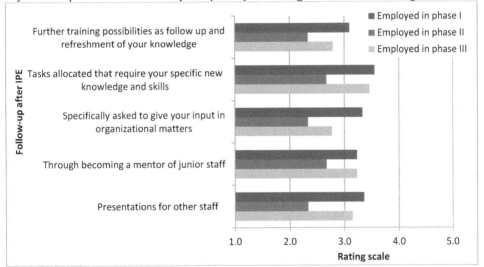

Figure 5.2. Follow up of IPE in the organisation

securing knowledge in a structured way. In an organic management system, greater importance is attached to knowledge coming from sources outside the organisation because knowledge from the sector can further the organisation's overall goals and performance (Burns and Stalker, 1961). A mechanistic management system tends to value internal knowledge more. The interviews[16] indicate that specialist knowledge, such as for functional positions, is not very highly valued in contrast to knowledge that benefits the administration and maintenance of the bureaucracy, as in line functions, which could also explain the modest follow-up of training and education.

---

[15] PD27, PD37, PD50, PD52, PD53, PD54, PD58
[16] PD32, PD39, PD49, PD58

In the current circumstances where DGWR is confronted with both a declining number of experienced staff, and at the same time major water resources management challenges such as climate change, follow-up of education and training is important, to accelerate the absorption of new knowledge in the organisation. Training and its follow up should ideally also be connected to a more comprehensive HRM system with a range of career paths and a subsequent training plan connected to these career paths. Currently, it is difficult to formulate demands for knowledge in a systematic way because of a lack of overview of where this knowledge is located in the organisation. To obtain knowledge required for decision-making, ministers and senior officials therefore rely on personal networks that limit the range of input (Sherlock, 2010).

Figure 5.3. KCD mechanisms that need a formal arrangement in each phase Rating scale: 1 = not at all, 3 = to some extent, 5 = extensively

5.3.4.2 KCD mechanisms

Figure 5.3 shows the formal KCD mechanisms in the organisation. It shows the low score for coordination with other departments and organisations. The interviews[17] confirm this picture. The organisational structure allows for vertical communication only. Opportunities for discussion and knowledge exchange with other stakeholders are not exploited in a formal way. The fact that survey and interview respondents unanimously state the low level of coordination shows the extent of the problem.

The use of KCD mechanisms other than formal education was investigated in the survey Figure 5.4 shows the informal KCD mechanisms that are used. Mentoring and coaching is, in contrast to the Dutch case study (Section 7.4.4), classified as an informal KCD mechanism, because in the Indonesian case the respondents indicated that mentoring was not arranged by the organisation.

Respondents score 'Learning by doing' (Figure5.4) rather high; especially in Phase III. Much learning happens in an unstructured fashion by trying things out. While this is an effective KCD tool in itself, the interviews also indicate that staff do not receive much personalised training in their work, so often learning by doing is their only choice.

Although formalisation and centralisation in the DGWR are dominant, many individual informal initiatives exist, partly because the formal means are not appropriate anymore to do the work. They are reflections of the past. Depending on the superior, there is room for staff to communicate their ideas and to take initiatives. This is confirmed by the interview data[18]. Informal knowledge exchange is a good development considering the speed with which tacit knowledge needs to be transferred before seniors retire, but the dependency on these individual informal initiatives to acquire new knowledge appears to be high.

A KCD mechanism that is mentioned several times, in particular by Phase I respondents is the mentoring relationship (Figure6.13). Knowledge transferred through a long process of mentoring is characterized by high viscosity, with the recipient gaining a significant amount of tacit knowledge, but only after a long period of time (Bhagat et al., 2002). High viscosity refers to the richness and retention of the knowledge transferred; the receiver will gain a tremendous amount of detailed and subtle knowledge over time (Davenport and Prusak, 1998). This takes observation, imitation, and practice (Nonaka and Takeuchi, 1995).

---

[17] PD27, PD32, PD33, PD35, PD36, PD38, PD39, PD42, PD49, PD50, PD56, PD58
[18] PD38, PD39, PD56, PD62

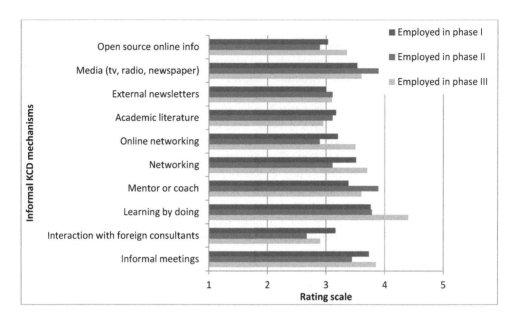

Figure 5.4. Informal KCD mechanisms in the DGWR in each phase. Rating scale: 1 = not at all, 3 = to some extent, 5 = extensively

Nonaka and Takeuchi (2011) further mention that among other measures, wise leaders make sure that apprenticeships and mentoring is organized, to cultivate practical wisdom in others. In slight contrast with Figure 5.4, in which Phase II respondents have rated this mechanism higher, the interview results[19] indicate that particularly all Phase I respondents stress that apprenticeships existed in Phase I, however were not institutionalized. I may conclude, however, that in Phase I tacit knowledge transfer was well taken care of. Knowledge exchange may need an organisational culture that at least to some extent is open for discussions, but this evidence shows that at least one directional knowledge transfer from senior to junior is possible in cultures with larger power distances. In Phase II, tacit knowledge transfer through apprenticeships received less attention. First of all, the zero-growth policy stopped the supply of graduates, thus gradually diminishing the need for apprenticeships, and secondly, the political climate rewarded loyalty and rent-seeking behaviour, rather than learning. In Phase III, apprenticeships were not re-introduced as a means of KCD. This may be a missed chance as in this phase the need for tacit knowledge transfers from senior to junior staff is of utmost importance, as the senior staff will retire within a few years and in addition, it is a means for KCD that is endogenous to the organisation, therefore presumably relatively easy to implement. It would also fit with the relatively mechanistic management system of the organisation as it focuses entirely on internal knowledge and fits well with cultural traits such as respect for senior staff and loyalty.

---

[19] PD36, PD39, PD40, PD46, PD48, PD49

5.3.4.3 Capacity for outsourcing

A KCD mechanism that was not tested in the survey but was shown to be important in the interviews[20] is outsourcing. Outsourcing involves the transfer of goods and services production, previously carried out internally, to an external provider (Domberger, 1998). It is a concept that fits the idea of an organic management system rather than a mechanistic system, as it requires acknowledgement that an outside party may be able to do certain work better than the outsourcing organisation itself and that it will improve the overall quality of the task of the organisation. Outsourcing is important for KCD because it has significant influence on knowledge coming in or leaving the organisation.

Interview respondents have stated that firstly, a standard needs to be set to be able to measure the quality of consultants work. Secondly, the tendering process needs to become more transparent to increase the quality: local consultants in Phases II and III are often retired MPW officials who were working in the MPW in Phase I. These consultants may sometimes be selected because of their ties to the DGWR instead of on a merit-basis. Thirdly, in Phase II, an increasing amount of technical specialist work was outsourced to local consultancy companies, gradually decreasing the expertise in the DGWR and MPW itself. The officials needed to supervise the contracted work are not available because the organisation has changed from consisting mostly of technical specialists to mostly administrators. All interview respondents of Phase I and resource persons have expressed concern over this and suggest that at least part of the technical work should be executed in force account.

5.4 CONCLUSION

I have investigated the influence of the organisational structure on KCD by looking at the elements that influence KCD specifically: the administrative divisions, the career system, training and other KCD mechanisms. Furthermore I have investigated the actual behaviour in relation to these elements. I have combined the elements of the formal organisational structure with the characteristics of mechanistic and organic management systems and have indicated where the DGWR needs to move to a more organic management system, to allow for more effective KCD in order to be better prepared for water management challenges in the country.

Three problems can be identified, related to the mechanistic tradition of the DGWR:

1. Until approximately 2020 a large gap will remain to exist in human resources, due to the zero-growth policy and the mechanistic prescription of the Ministry of State Apparatus on the quantity of new staff to be recruited. Few mechanisms are available to acquire and share knowledge and capacity in the organisation, partly because of the administrative divisions that limit lateral communication. The few KCD mechanisms that exist are not institutionalised.
2. Although training is in place to provide competence in routine tasks, the HRM system is insufficiently prepared to provide specialist knowledge to cope with challenges such as climate change.

---

[20] PD32, PD36, PD39, PD40, PD47, PD48, PD49, PD53, PD54, PD57

I recommend the following measures:

1. Decentralise responsibility to the lowest level possible. Staff would feel more ownership for their work, and see where they fit in the wheelwork of the organisation. Knowledge would not be solely at the top of the hierarchy, but may be located anywhere in the organisation.
2. Facilitate and institutionalise selected KCD mechanisms. The lack of follow-up of training and especially of IPE is a missed chance for KCD, especially as IPE will remain to be a relative important mechanism to acquire knowledge.

    Increased interaction between staff would result also in recruitment, training and career paths that are better connected to the knowledge needs of the organisation and individual staff.
3. The apprenticeship is an excellent way to gain and transfer tacit knowledge, and a form of succession planning as well. It is furthermore a good fit with the current relatively mechanistic management system of the organisation, as it focused on internal knowledge and fits well with cultural traits.
4. Outsourcing could be a way to make use of external knowledge; however, technical capacity within the organisation is needed to supervise the quality of the work. Administrative capacity is required to turn outsourcing into a transparent process, based on merit-based competition between consultants.

The relation between formal organisational structure and the rules in use will be further investigated in the Discussion.

6 Competence formation and post-graduate education in the DGWR

6.1 INTRODUCTION[21]

Many of today's water challenges are so complex in nature that they require the involvement of multiple disciplines and the collaboration of several organisations (Alaerts and Dickinson, 2007; Bourget, 2008; Loucks, 2008; Nash et al., 1990; Wagener et al., 2007; Wagener et al., 2010). It can safely be predicted that water problems are likely to become more serious, certainly in developing country contexts, and that water resources use, management and governance will continue to be a politically contested terrain (Mollinga, 2009).

In facing the challenges, water professionals are needed with specialisations in particular disciplines, and conversance with other relevant disciplines. An engineer should not only have an operational knowledge of theories and principles of mathematics, physics, chemistry, engineering economics and statistics, but also have an understanding of behavioural processes, systems analysis and computer modelling, laws and regulations, history, sociology and ethics (Loucks, 2008). Water professionals need to be able to cross boundaries: disciplinary boundaries, but also boundaries in society.

In this chapter I build further on the conceptualisations of competence presented in the Theory Chapter, comprising three aggregate competences, namely for technical issues, management and governance, together with a meta-competence for continuous learning and innovation. The latter is a prerequisite for obtaining and improving the other aggregate competences. The aggregate competences are further operationalised in a cognitive - explicit, a cognitive - tacit, functional, personal, and values or ethical component (Sultana, 2009). The aggregate competences are subsequently organised into various T-shaped competence profiles with the vertical bar representing a substantive specialisation, often disciplinary, and the horizontal bar representing competence in adjacent disciplines, a meta-competence for continuous learning and innovation, to enable the professional to act and collaborate across boundaries.

The framework is presented in Section 6.3. Section 6.4 elaborates on the results and discusses the competences acquired in post-graduate education in relation to the needs as perceived by the Indonesian water professionals. I conclude in Section 6.5 with suggestions for research and practical recommendations for Human Resource Management and for education.

6.2 SUBCOMPONENTS OF COMPETENCE AND THE T-SHAPED COMPETENCE PROFILE

In Chapter 2 I described the overall adapted KCD conceptual model, which included a description of three aggregate competences and one meta-competence that I deem important for the public water sector. For the organisational and institutional level the first level of operationalisation suffices because for these levels it is mainly important that the organisation has the appropriate mechanisms in place to facilitate KCD, such as an HRM policy and implementation and furthermore that it has a clear image available of what the appropriate competence mix is at the organisational level. Which type of KCD

---

[21] This chapter is adapted from an article published by Kaspersma et al. Kaspersma, J. M., Alaerts, G. J., and Slinger, J. H.: Competence formation and post-graduate education in the public water sector in Indonesia, Hydrol. Earth Syst. Sci., 16, 2379-2392, 10.5194/hess-16-2379-2012, 2012a. in Hydrology and Earth System Sciences.

mechanisms are required can be based on the analysis of the individual competence needs. For the individual level I build further on the initial conceptualisations and add complementary theory to the adapted KCD conceptual model.

Knowledge and competence. There is substantial debate in the literature concerning the concept of 'competence', and it is impossible to identify a conclusive theory, or to arrive at a definition capable of accommodating and reconciling the multitude of ways in which the term is used (Delamare Le Deist and Winterton, 2005). However, approaches that were developed relatively independently, in the United States (McClelland, 1976, 1998; Prahalad and Hamel, 1993; White, 1959), the United Kingdom (Cheetham and Chivers, 2005), and Germany and France (Bohlinger, 2007/2008) have given way to frameworks that see competence as a multi-dimensional holistic concept (Delamare Le Deist and Winterton, 2005), including (i) a cognitive-explicit component, that involves the use of objective and replicable theory and concepts, as well as (ii) an informal cognitive-tacit component, which is gained experientially; (iii) a functional component (skills or 'know-how'), i.e., those things that a person should be able to apply when functioning in a given area of work, learning or social activity; (iv.a) a personal component, involving attitudes and knowing how to conduct oneself in a specific situation; and (iv.b) a values or ethical component involving the possession of certain personal and professional values (Sultana, 2009). I treat the personal and ethical component as one, as personal attitude is largely the consequence of one's norms and values.

The concepts of knowledge and competence have different intellectual roots but they share the same ingredients. In Section 2.2.2 I introduced four components of knowledge: Information, experience and understanding, skills and attitude.

Information equals the explicit part of cognitive competence, skills are similar to functional competence, experience is similar to the tacit part of cognitive competence, and attitude is comparable to personal and ethical/ values competence. Furthermore, in the context of international development, these two concepts are also intimately related to that of 'capacity', which refers to the ability of organisations or individuals to be effective in their endeavours (Alaerts and Kaspersma, 2009).In this chapter I make the pragmatic choice to work with the concept of competence instead of knowledge, as most of the literature related to education refers to competence instead of knowledge.

An overview of the three aggregate competences and the meta-competence for continuous learning and innovation is provided in Table 6.1 together with the four components and practical examples. The competence for continuous learning and innovation is termed a meta-competence because it exists beyond the other competences and enables individuals to monitor and develop the other competences (Cheetham and Chivers, 2005). I have displayed the meta-competence for continuous learning and innovation as an umbrella over the other three competences because it is a prerequisite for every professional, whether he or she specialises in a technical, management or governance subject. The professionals working in these organisations will need aspects of the technical, management and governance competence to a certain degree, together with the competence for continuous learning and innovation. The mere possession of knowledge and expertise in the professional's own field is in most cases no longer sufficient. It is now necessary to have a basic knowledge—though not necessarily an operational grasp—of adjacent and connecting fields in order to work in multidisciplinary ways and be a good discussion and collaboration partner, both within and outside the organisation (Oskam, 2009). The combination of essential aggregate competences can be represented visually by a T-shape. A professional with a T-shaped competence profile has specialist knowledge in their own field (the vertical leg of the T), plus a broad knowledge base with elementary knowledge or insight in adjacent water fields or more general disciplines such as business administration (the horizontal leg of the T) as well as soft-skills enabling him or her to communicate with other disciplines (Oskam, 2009; Mollinga, 2009; Uhlenbrook and De Jong, 2012). In Figure 6.1 I provide a possible profile for a technical water specialist. For example, a hydrologist will need cognitive explicit competence in mathematics, physics, hydrology, water resources systems, and similar 'basic' disciplines.

Table 6.1. Three aggregate competences, and one meta-competence for continuous learning and innovation, for professionals in the water sector, based on Alaerts (2009) and Kaspersma (2009).

Aggr. Competence:	Meta-competence for continuous learning and innovation			
	Cognitive - explicit component	Cognitive - tacit component	Functional component	Personal /ethical/ value component
	Knowledge about learning and learning styles	Experience with and awareness of your learning style	Critical thinking, self-discipline, ability to concentrate	Availability for training and education in new knowledge, readiness to critically reflect on one's own performance, desire to 'keep learning', creativity, self-confidence, non-individualistic attitude
Technical	Regularly updated technical knowledge	Understanding of the broader technical context, application insight, intuitive understanding	Design skills, modelling skills	Conscious choices on where and how to build infrastructure
Management	Regularly updated knowledge about management	Understanding of broader organisational context, Application insight, intuitive understanding	Project management skills, financial management skills, people mgmt, negotiation Mentoring, Ability to 'deliver', Leadership	Willingness to involve staff in decision-making, Knowledge sharing attitude
Governance	Regularly updated knowledge on governance, such as participation stakeholder involvement	Ability to apply inclusiveness, Understanding of procedures and institutional structures, Understanding of political consensus building Application insight, intuitive understanding Ability to cross disciplinary boundaries	Policy formulation skills, Working in a participative manner	Achieving ethical objectives: non-corruption, transparency, etc. Willingness to cross disciplinary boundaries (Mollinga, 2009)

Furthermore, he/she will need a broad understanding of the technical context in which he/she is working (cognitive tacit competence), for instance the fields of river or flood management. He/she also needs functional competence or skills such as hydrological modelling. These all belong to the technical aggregate competence. In terms of the aggregate management competence, a hydrologist will at least need understanding of the organisational context in which he/she works such as an the operational rules, regulatory constraints or research establishment procedures (cognitive tacit component), some project management skills, mentoring skills for junior staff (functional competence) and a knowledge sharing attitude and commitment to the job (personal/ethical/value component). Similarly he/she will require a willingness and ability to effectively work in teams across disciplinary boundaries for higher goals (aggregate governance competence). Ideally, the hydrologist should also develop his/her learning style, ability to think critically, and openness for continuous learning (meta-competence for continuous learning and innovation). Depending on the specialisation and the water sector context, other components of the technical, management and governance competence may be important. Still, the competence mix ('capacity') of the organisation is what matters most, with some staff being highly specialized and mono-disciplinary (long and narrow T's), but with a growing majority equipped with shallower yet broader T's. Those with broader T's serve to enhance team cohesion and overall effectiveness. This chapter does not discuss the specific competences needed for given situations nor the optimum ratio of specialisation to breadth for an individual. Similarly, I do not address the mix of T's required for an organisation to enhance its effectiveness. Instead, I note that different mechanisms exist to develop these competences and focus on the role of one particular mechanism in studying competence formation.

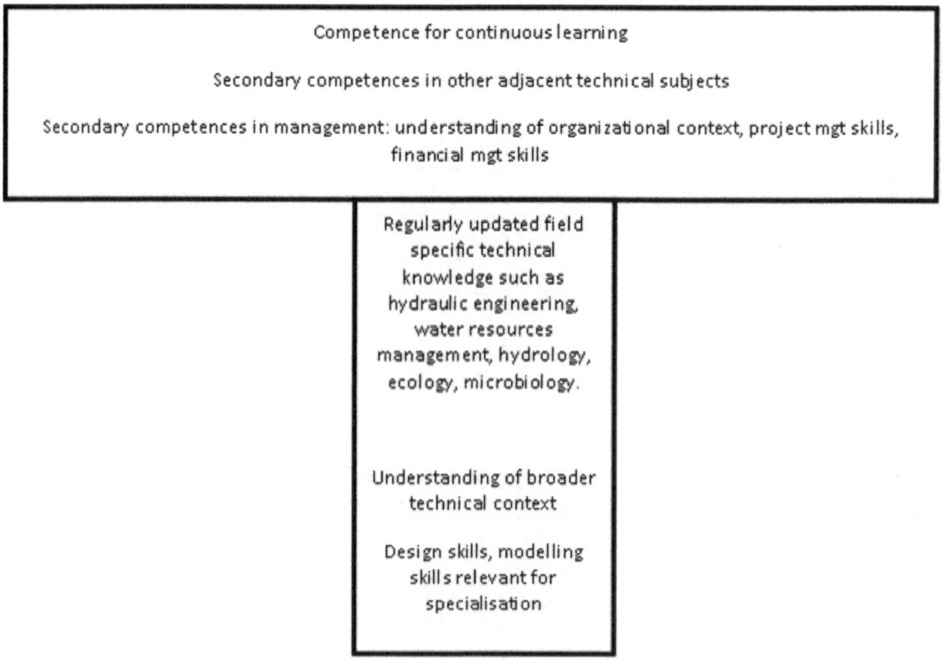

Figure 6.1. Hypothetical T-shaped competence profile for a technical water specialist

Typical instruments for competence formation are formal education and training, which are suitable for acquiring the cognitive - explicit component, whereas the tacit component can best be transferred through one-on-one interaction between junior and senior, apprentice and mentor. Also networks – both formal and informal associations and 'communities of practice' – are important mechanisms for professional improvement for many water professionals (Alaerts and Kaspersma, 2009). Depending on socio-economic conditions in a country, some mechanisms are more prominent than others. In less developed countries, post-graduate education often fulfils a key role in generating competence and accessing specialized knowledge, as other mechanisms to access this type of knowledge (professional associations, high-quality seminars, etc.) are less abundantly available. Finally, the perception of what competences are most needed, and how they should be acquired and enhanced, depends on policies and procedures in the sectoral institutions, and in those of the 'enabling environment' (Alaerts and Kaspersma, 2009). Competence valuation and acquisition, therefore, depend strongly on external factors of a policy, administrative and cultural nature.

6.3   RESULTS AND DISCUSSION

6.3.1   Three aggregate competences and the learning meta-competence

The aggregate management and governance competences receive a much higher appreciation than might be expected considering the strong technical orientation and history of the DGWR and the respondents (Figure 6.2 and 6.3). It should be noted that the aggregate competences and the learning meta-competence are mutually non-exclusive response options, allowing higher scores on more than one option. Both LPE and IPE respondents attach lower weight to the aggregate technical competence (Figure 6.2). These results are at odds with the outcomes of the interviews[22], and also with the historical and current incentive system and the expectations at the DGWR, which both tend to stress the aggregate technical competence. Across the board, the interviews stress that professional staff at the DGWR know that the other aggregate competences should become part of the overall set of aggregate competence of the DGWR. However, in practice, the orientation has remained overwhelmingly technical thus far.

Respondents with IPE experience tend to rate the need for competence higher than the respondents with LPE experience (Figure 6.2). It seems they have stronger opinions about the need for, and value of, competence than LPE respondents. The interviews[23] confirm the general value attributed to learning, but indicate, in addition, that the need for learning is especially highly appreciated by IPE respondents, and is viewed as key to personal and organisational advancement, in an environment where the number of individuals with a post-graduate education is still limited.

---

[22] PD27, PD30, PD31, PD32, PD33, PD34, PD39, PD41, PD42, PD50, PD51, PD58, PD62, PD64, PD65, PD66
[23] PD27, PD30, PD31, PD37, PD38, PD39, PD41, PD46, PD50, PD56, PD57, PD58, PD61, PD62

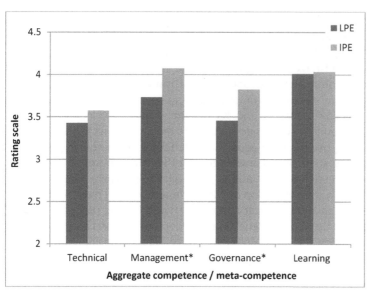

Figure 6.2. Aggregate competences and meta-competence perceived to be required for daily tasks in the DGWR as rated by LPE alumni and IPE alumni, rating scale from 1 to 5, where 1 = not at all, 3 = to some extent, and 5 = extensively

Respondents from Phases II and III all tend to under-value the need for aggregate technical competence (Figure 6.3). This contrasts starkly with the reality of low levels of technical competence at the DGWR during Phases II and III. Furthermore, the meta-competence for continuous learning and innovation is perceived as highly necessary especially by respondents from Phases I and III.

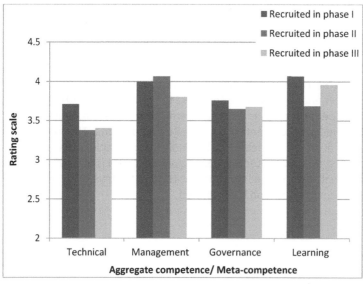

Figure 6.3. Aggregate competences and meta-competence perceived to be required for daily tasks in the DGWR, in Phase I, II and III, rating scale from 1 to 5, where 1 = not at all, 3 = to some extent, and 5 = extensively

Although the differences are statistically not significant, the interview outcomes underscore this result. In Phase I, DGWR was characterised by a strong *esprit de corps*, and a focus on technical competence and development, also of skills, whilst Phase III respondents are much younger and are likely to recognize the necessity to improve on their specialisation, which is likely less technical[24].

Most of the LPE programs are perceived to not yet strongly reflect the paradigm shift from 'construction' to water resources management and to a new set of resulting competences (Figure 6.4) (Ministry of Public Works - Directorate General of Water Resources, 2010). This observation is confirmed by both the survey and the interviews. During the interviews, all respondents confirmed the technical orientation of locally-based post-graduate education, but they add that this is the most directly appreciated educational background in the DGWR.

For IPE respondents, the interviews and the survey show the same trend (Figs. 6.4 and 6.5) with high scores for the aggregate technical competence, and the meta-competence for continuous learning and innovation, compared to the other aggregate competences. Both interview[25] and survey respondents report that the IPE experience was particularly effective in acquiring specialised new skills such as computational techniques, as well as for the cognitive - tacit component of competence. Still, the idea that IPE might emphasise the non-technical competences more, seems not to be borne out. Furthermore, it is surprising that the aggregate management and governance competences were scored higher by LPE than IPE respondents in the survey. This does not necessarily mean that LPE is more effective than IPE in absolute terms, as respondents cannot compare the effectiveness of IPE and LPE, because each respondent has had only one such experience. In fact, the analysis of documentation and the interviews, in which the interviewer as outsider is able to assess both types of education, do not the support the perception of the respondents in this case.

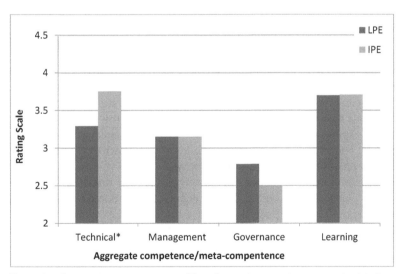

Figure 6.4. Aggregate competences and learning meta-competence perceived to be acquired during LPE and IPE, rating scale from 1 to 5, where 1 = not at all, 3 = to some extent, and 5 = extensively

---

[24]PD27, PD29, PD30, PD32, PD33, PD34, PD38, PD39, PD40, PD51, PD63, PD65
[25] PD27, PD30, PD32, PD41, PD51, PD54, PD62

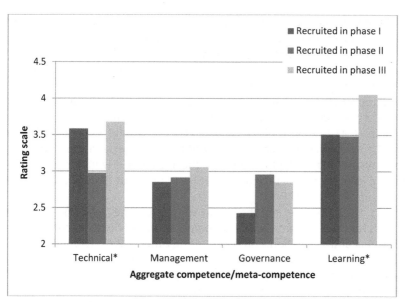

Figure 6.5. Aggregate competences and meta-competence perceived to be acquired during LPE or IPE in Phase I, II and III, rating scale from 1 to 5, where 1 = not at all, 3 = to some extent, and 5 = extensively

Phase I respondents plausibly emphasise their acquisition of the aggregate technical competence. Indeed water management concepts such as IWRM and water governance were not widespread, let alone incorporated in curricula, in that period (Figure 6.5). We would expect this to be similar for Phase II, but the survey results reflect a lower level perceived need and formation of technical competence. A much higher appreciation for aggregate technical competence is again reflected among the young respondents from Phase III, whereas in Phase III there are more educational programmes available with an IWRM orientation. Moreover, in Phase III, apart from the increasing attention for aggregate competences other than the technical, there is a new need for increasing the technical competence after a long period of erosion of this expertise, first in the aftermath of Phase II and then during the financial crisis at the onset of Phase III. Therefore, strictly technical subjects will remain a popular choice for students enrolling in an IPE or LPE. The lower score in Phase II could be a reflection of the perceived higher need for loyalty to the regime to the detriment of technical competence, as described in Chapter 4, Section 4.3.2 and in the interviews[26]. The need to obtain technical competence through education may have been a relatively low priority.

Both the aggregate governance and management competence are rated low in Figure 6.5. For Phase I, this is plausible considering the dominant construction paradigm in the period. A low rating might also be expected for Phase II, but the survey results do not bear this out. The interview results[27] and literature confirm that in Phase II, despite attempts to introduce the concept of IWRM and the IOMP, the organisational focus was still heavily on construction and infrastructure development. The IPE interviewees employed in Phase III furthermore indicated[28] that they consciously chose for a water

[26] PD32, PD33, PD37, PD38
[27] PD31, PD32, PD33, PD38, PD54
[28] PD27, PD62

governance or water management orientation, stating that they had acquired the aggregate technical competence during their BSc degree or within the organisation itself. However, having an aggregate technical competence is a core priority of the MPW, and for post-graduate education a technical subject may still have the preference of many staff.

The difference between the aggregate management and governance competences as acquired from post-graduate education and the aggregate competence as required for daily tasks is relatively large, when we compare Figs. 6.2 and 6.3 with Figs. 6.4 and 6.5. For the aggregate management competence, the need in daily work is explained by the fact that work at the DGWR consists for a large part of administrative tasks, which forms part of the aggregate management competence. This is mentioned in all interviews. This finding suggests that a discrepancy exists between what is needed in daily work and the orientation people choose in their post-graduate education.

For the aggregate governance competence, the IPE respondents score the extent to which an aggregate competence is needed in daily work higher, whereas the LPE respondents surprisingly score the extent to which they obtained this aggregate competence in school higher. The latter is not confirmed by the interviews. Interviews with resource persons and LPE as well as IPE respondents raise the issue of the gap between the aggregate governance competence in the organisation and the extent to which it is obtained during an LPE or IPE[29]. We can conclude from these results that for the aggregate management and governance competences an apparent imbalance exists between the aggregate competences required in daily work vs. what is obtained during post-graduate water education, be it local or international. This insight is confirmed by both the survey and the interviews.

---

[29] PD27, PD30, PD32, PD38, PD50, PD52, PD58, PD61

6.3.2 Technical competences

The aggregate technical competence consists of four components. The cognitive - explicit component, e.g., specialised theoretical knowledge; the cognitive - tacit component, e.g., a broader understanding of the technical context; and a functional component, e.g. design or modelling skills and a personal/ethical/values component, e.g. participatory design (Figure 6.6).

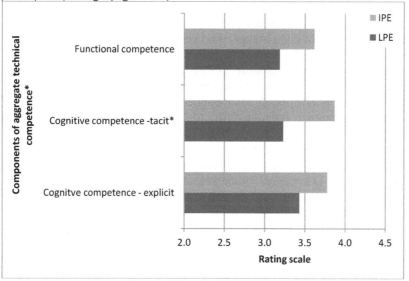

Figure 6.6. Components of aggregate technical competence perceived to be acquired during LPE and IPE, rating scale from 1 to 5, where 1 = not at all, 3 = to some extent, and 5 = extensively

The results for LPE and IPE respondents show higher scores for IPE, for each component. The LPE programmes for water management show a slight preference for the cognitive - explicit component of the aggregate technical competence. This is also confirmed in the interviews with LPE respondents[30]. For IPE respondents, the cognitive - tacit component is perceived to be higher than the others. The interviews confirm that IPE pays more attention to the cognitive - tacit component by, for example, creating broader understanding of the technical context, and by technical problem solving[31]. Further comparative analysis of the specific sub-competences did not reveal information additional to that already discussed in Section 6.3.1.

6.3.3 Management competences

A need is perceived to exist for each component of the aggregate management competence (Figure 6.7). The need for the cognitive-tacit component is valued slightly more than the others. The cognitive - tacit component includes competences such as a broader understanding of how organisations and their staff are managed. Phase III respondents rated the need in daily work of the cognitive-explicit component (e.g. theoretical knowledge about organisational management and people management) lower than respondents of the other two phases; the reason for this low rating is unclear.

---

[30] PD50, PD52, PD58, PD61
[31] PD27, PD36, PD38, PD51, PD54, PD62

This component of the competence, though essential, would automatically also become tacit, i.e., it is internalized when people start applying it in work situations they encounter. This may explain why the cognitive - tacit component is rated higher than the cognitive – explicit component.

When the management competence acquired during post-graduate water education is broken down into components (Figure 6.8), the rating of the explicit component is much lower than the rating of the other components. Phase III respondents emphasised[32] that the education seemed to favour the functional component, constituted by competences such as written communication skills and project management; and the personal/ethical/values component, such as the willingness to share knowledge. The components, and especially the cognitive explicit component, appear insufficiently addressed in the curricula of both IPE and LPE.

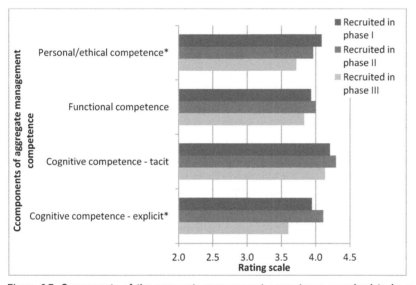

Figure 6.7. Components of the aggregate management competence perceived to be required in daily work in Phase I, II and III, rating scale from 1 to 5, where 1 = not at all, 3 = to some extent, and 5 = extensively

---

[32] PD27, PD30, PD36, PD38, PD56, PD62

Figure 6.8. Components of the aggregate management competence perceived to be acquired during LPE and IPE in Phase I, II and III, rating scale from 1 to 5, where 1 = not at all, 3 = to some extent, and 5 = extensively

6.3.4 Governance competences

As mentioned in Section 6.3.1, there is an imbalance apparent between the high rating of the aggregate governance competence required in daily work versus the markedly low rating for its acquisition during post-graduate water education, be it local or international (Figs. 6.9 and 6.10). The cognitive - tacit component, however, such as for example the understanding of political consensus building, is rated low in both figures, especially by Phase I respondents. This can be explained by the fact that in Phase I a strong *esprit de corps* existed because of the shared goal of the development of the nation, albeit under a highly centralised administrative and political system. There was only limited space for political consensus building as decisions were made by the top management, and these were the years where the power and successes of the MPW were at their peak. Therefore there was no perceived need for governance competence in daily work, and no reason to acquire this component in education.

The acquisition of the personal/ethical component, e.g. the ability to achieve ethical objectives, is scored higher by respondents from Phase III (Figure 6.10). Considering the more open atmosphere in Phase III, with more attention for accountability and transparency in the DGWR, it is to be expected that the personal/ethical component scores higher, but this would naturally also be expected in Figure 6.9. The results from the interviews[33] indicate that the younger generation of Phase III tends to appreciate the personal/ethical/values component, and the organisation also provides more space for this.

---

[33] PD27, PD50, PD54, PD58, PD62

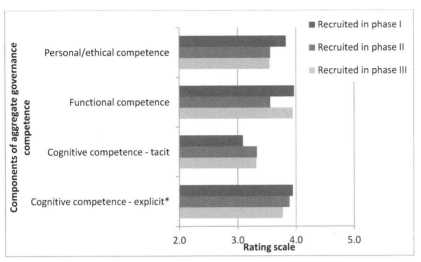

Figure 6.9. Components of aggregate governance competence perceived to be required in daily work, in Phase I, II and III, rating scale from 1 to 5, where 1 = not at all, 3 = to some extent, and 5 = extensively

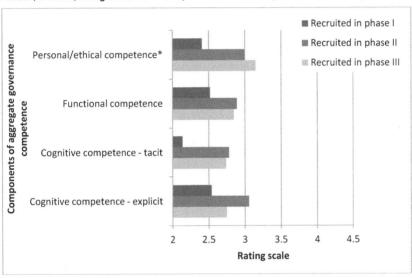

Figure 6.10. Components of the aggregate governance competence perceived to be acquired from LPE and IPE, in Phase I, II and III, rating scale from 1 to 5, where 1 = not at all, 3 = to some extent, and 5 = extensively

Multidisciplinarity. The survey results reveal that multidisciplinary thinking acquired during education is rated significantly lower for LPE than for IPE, although the score for LPE is not extremely low. In the interviews, though, no evidence is found pointing to multidisciplinary working skills as obtained in LPE. In IPE the situation is slightly different. Even though aggregate competences such as management and governance are not taught sufficiently intensively according to the survey and interviews, and consequently connections between these topics and technical disciplines are weak, interview respondents stated that they were encouraged to look over the boundaries of their

103

discipline and learn how to communicate with colleagues from other disciplines[34]. However, for both IPE and LPE opportunities for improvement exist.

6.3.5 Meta-competence for continuous learning and innovation

Graduates of both IPE and LPE rate the need for the meta-competence for continuous learning and innovation in daily work very highly in comparison with the other competences (Figs. 6.2 and 6.3), and score the formation of this competence during education slightly higher than the other competences (Figs. 6.4 and 6.5). Phase I respondents pointed to the DGWR policies during Phase I, which stressed the continuous learning concept and had in place mechanisms to rotate staff, provide mentorship and junior-senior learning arrangements, and to the careful selection of IPE candidates (during this period, no LPE was on offer yet). On the other hand, Phase II respondents rate this need decidedly lower, probably because in Phase II competence and learning were sometimes subordinated to loyalty, when it came to incentives and careers. Yet, also in this phase DGWR had suitable policies in place albeit possibly less well enforced. The cognitive - explicit component (for example, intercultural sensitivity) of the meta-competence is rated lowest of the four components in terms of its need in daily work (Figure 6.11) as well as the extent to which it is acquired in post-graduate education (Figure 6.12). The perceived low need in daily work is remarkable as intercultural sensitivity is important in the Indonesian context, considering the diverse ethnical backgrounds of the staff of the DGWR.

Explicit attention for this meta-competence is only marginally present in the water curricula of LPE and IPE, and this is confirmed by both the survey and the interviews. This means that the cognitive - explicit component of the meta-competence is not taught. The other components of the meta-competence, i.e., the cognitive – tacit (for example, the ability to reflect on oneself); the functional (such as critical thinking); the personal/ethical/values (for example, creativity) are implicit in the curricula; they can, to some extent, be acquired while working on the other aggregate competences.

The interview outcomes, contradicting the survey results, confirm the acquisition of the meta-competence for continuous learning and innovation from IPE education only[35]. For example, all Phase III interview respondents with an LPE stated that the meta-competence for learning is a personality trait and exists independent of what one learns in school; they furthermore thought that they did not acquire this competence in school. IPE alumni, on the other hand, unanimously stated that acquiring the meta-competence for continuous learning and innovation while abroad was synonymous with the fundamental attitude changes associated with the IPE experience.

---

[34] PD27, PD36, PD38, PD41, PD54, PD62
[35] PD27, PD36, PD38, PD41, PD54, PD62

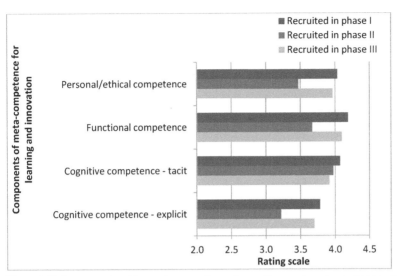

Figure 6.11. Components of the meta-competence for continuous learning and innovation perceived to be required in daily tasks, in Phase I, II and III, rating scale from 1 to 5, where 1 = not at all, 3 = to some extent, and 5 = extensively

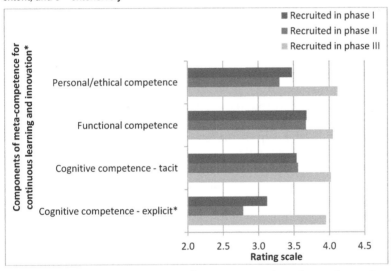

Figure 6.12. Perception of components of the meta-competence for continuous learning and innovation perceived to be acquired from IPE and LPE, in Phase I, II and III, rating scale from 1 to 5, where 1 = not at all, 3 = to some extent, and 5 = extensively

All IPE interviewees, together with 13 out of 14 resource persons confirmed the hypothesis that the most important aspect of IPE is acquiring the meta-competence for continuous learning and innovation, and the attitude change that happens simultaneously. This has been termed socialisation of tacit knowledge (Nonaka and Takeuchi, 1995). The aspects of education that catalyze the socialisation are summarised in Table 6.2.

105

Table 6.2. Most important aspects of international education, as perceived by interview respondents

Aspect	How many times mentioned in interviews
Learning how to learn, encouragement of curiosity	10
Intercultural sensitivity	7
Being exposed to new concepts	10
Discussions with peers	7
Discussion with professors	5
International networking	7
Encouragement of critical thinking	10
Working in an interdisciplinary fashion	6
Team building	6
Being away from home/living in a different country	5
Working independently	4
Exercises for self-reflection	1
Learning to plan ahead	6
Different working culture	8
Connecting theory to practice	6

Learning is encouraged by group work, role plays, debates, and through a working culture with smaller power distances in human relations, facilitating easier interaction between professors and students. Typical statements from the interviews confirm this picture: *'The very fact that you are in a different country changes the experience. You are told things you wouldn't accept from someone in your home country[36].'*, and *'Receiving critical comments from teachers was difficult. They would tell you that you need to do certain things differently, but not how you should do it. You have to find out yourself but they help you getting started[37]. '*

The one resource person[38] who doubted the attitude change among alumni from international post-graduate water education suggested that MPW staff should learn together with NGO staff, civil society representatives and university staff. Indeed, the large majority of participants in international programmes have a government background. Creating a more heterogeneous mix of backgrounds and opinions in educational programmes would help increase the learning opportunities.

6.4 DISCUSSION

An adapted KCD conceptual framework was introduced, comprising three aggregate competences for technical issues, management and governance, and a meta-competence for continuous learning and innovation for the water sector. The framework was tested in a case study on post-graduate water education for staff in the DGWR of the MPW in Indonesia. Though DGWR and the sector professionals have a firmly 'technical' default orientation, both the surveys and interviews reveal a strong perceived need for the other competences: primarily the meta-competence to learn and innovate, as well as the aggregate competence of management. The need for the aggregate competence of governance is systematically rated lower; however, it is still regarded as reasonably

---

[36] PD27
[37] PD62
[38] PD42

important. A discrepancy appears to exist between the competences that staff perceive as needed in daily work, compared to those that can be acquired during post-graduate water education. Respondents also indicated that the DGWR needs a wider palette of competences beyond the strictly technical ones, but these competences seem not to be acquired intensively from the post-graduate education, even though some international courses do offer them. Respondents also mentioned that eventually a technical education is the most directly appreciated educational background in the DGWR. This apparent contradiction can be explained by a possible path-dependency that influences people to choose a familiar subject, which is already appreciated in their organisation. As mentioned earlier, the interviews indicate that many people are aware that the DGWR needs expertise in fields other than water engineering but in practice, engineering is still the basic and, perhaps too often, singularly appreciated educational background.

Thus, it can be assumed that the DGWR is still located in a primarily mono-disciplinary technical paradigm. This situation is unavoidable given the current need to rebuild and expand the existing infrastructure and the added burden of staff shortage. It is likely that in the coming decade this discrepancy will increasingly call for a new human resource policy at DGWR. It might be expected that international post-graduate education would have helped in addressing the need for the aggregate management and governance competences, but this has not been the case, possibly because respondents hail from three different phases in the development of the water sector and international courses did not then offer such competence building prior to, say, 1990.

The LPE in Indonesia has a strongly cognitive - explicit and technical orientation. With this set of knowledge, a part of the vertical bar of the T-shape is filled. The aggregate technical competence is not addressed completely, the cognitive - tacit component (understanding, practical experience) and concomitant adapting attitude, tend to receive relatively less attention. IPE programmes, on the other hand, have a relatively strong technical focus as well, but, in particular in Phase III, paid more attention to the cognitive - tacit component, in addition to the cognitive-explicit component. Amongst educationalists, the cognitive - explicit component is perceived as fundamental to post-graduate education, and is then internalised to become tacit, when it is combined with personal experience gained on-the-job.

In both LPE and IPE, the aggregate competences for management as well as governance are reportedly addressed modestly, if at all. For the aggregate management competence, the explanation may be that people expect to obtain the management competence through on-the-job training. Management training is indeed available for staff of DGWR, although this type of training has a strong focus on the functional component of the competence, i.e. skills, only.

The ability to think and work in a multidisciplinary fashion is an aspect of the aggregate competence for governance that merits special attention. The documentation on existing local water curricula shows that in most water-related MSc curricula the connections between the different disciplines are made insufficiently, and multidisciplinary thinking is limited to the technical disciplines.

With little competence in these fields, it is difficult for professionals to communicate and collaborate effectively in multidisciplinary settings and in the water policy and political discussions that have characterized the Indonesian water sector since the onset of Phase III. As a result, the horizontal bar of the T-shaped profile remains weakly developed. In IPE, this is partially compensated for by the (implicit) attention for continuous learning and innovation. The participants are in a totally different habitat for a substantial period and in the assignments they have the chance to mingle with different cultures and experience different thinking patterns and work attitudes, and they are encouraged to discuss with lecturers and professors and ask questions, which is a new experience for many participants. The exposure to a different culture and learning format is reported as fundamentally formative.

6.5 CONCLUDING REMARKS

In regard to both IPE and LPE, the responses in this study cover a substantial period and the study is not able to make conclusive statements about the performance of current post-graduate water education. Since 2000, the Indonesian water curricula in particular are becoming more responsive to the new demands of the sector. However, project-based learning could be used more extensively to train the multidisciplinary aspects necessary for enhancing management competence. To enhance the continuous learning and innovation competence educational programmes could include teaching on learning styles (the cognitive explicit component but also the functional and personal component) so that people are aware of their own learning style and can use this personal knowledge to help them in further competence formation. However, we have also indicated that eventually engineering is still the most directly appreciated educational background in the DGWR and these findings could well pertain to the water sector in other developing and emerging economies.

The competence framework cannot resolve decisions on the details of the required T-shaped competence profiles and skills mixes for specific water specialisations of individuals and organisations (e.g., on the relative importance of 'broad' competence versus 'in-depth' expertise) but it can help in outlining requirements that subsequently help guide MSc-level water curricula improvements. It could be useful to design a number of modules in an educational programme that focus solely on the horizontal bar of the T-shaped competence profile. Students should choose from a number of subjects outside their specialisation that should help them connect to other disciplines. Furthermore, it would be beneficial for the educational institute, the student and the home organisation of the participant if there was more exchange about the needs of the home organisation and its ability to work with the new knowledge. The student can then prepare his or her return more thoroughly.

Further, the methodological differentiation of respondents as a function of their LPE or IPE experience, and of the administrative and political context in the country and sector, has proven useful in generating detailed insights into development of competence formation in the Indonesian water sector, over a long span of time against the evolving economic, administrative and political contexts. However, the combined utilisation of guided surveys, semi-structured interviews and in-depth analysis of reports and policy papers proved essential for accurate and meaningful interpretation; survey results alone were often insufficient and difficult to interpret. Importantly, both the interviews and surveys reflect the perceptions of individuals. However many respondents have had only one and a very personal experience with LPE or IPE and their perceptions are not necessary mutually compatible. The presence of the interviewer/analyst particularly in the guided surveys offered the possibility to set robust assessment benchmarks and enhanced the quality of the data and its interpretation.

In view of the dominant position of the DGWR in the water sector within Indonesia and the care taken to span the different phases of water development as well as to interview water professionals external to the ministry, the findings of this research are considered representative for the water sector in Indonesia. Indeed, the findings could well be valid in the water sector in other developing and emerging economies. Further research aimed at determining the T-shaped competence profiles required for specific water specialisations (e.g., the relative importance of 'broad' competence versus 'in-depth' expertise) and the skills mixes required by water organisations is necessary, and could guide improvements in masters' level water curricula and the selection of IPE training by organisations.

7 'God created the Dutch, and the Dutch created the Netherlands...' - KCD in the Rijkswaterstaat of the Netherlands

7.1   INTRODUCTION

The Netherlands is viewed by outsiders as a country with a well developed water sector and a centuries old tradition of successful water management. Land reclamation, protection against river and sea floods and the generally effective treatment of wastewater are counted as national strengths spanning diverse aspects of water management. The Dutch word 'polder' has been incorporated in many languages. The fact that the apocryphal story of Hansje Brinker, who put his finger in the dike to stem a leak and so averted a flood, underscores this reputation rather than distract from it. In short, the Dutch created the Netherlands.

Documentation indicates (Karstens et al., 2011; Netherlands Water Partnership, 2011) that the water sector is currently still a prosperous sector with many different actors, producing innovative solutions to a variety of modern-day water problems, such as cost-effective removal technology of nutrients from domestic sewage, the construction of storm surge barriers and coastal protection, and the application of aquatic ecology to improve the water quality and biodiversity of surface waters, trying to balance different stakeholders and interests.

In order to create a context for the study of KCD in international development, it is interesting to investigate how KCD functions in a society with a more mature sector capacity and less financial constraints, hence where assumedly many KCD mechanisms are available and put to effective use. For example, many experts are active member of professional water associations such as the International Water Association (IWA), represented in the Netherlands by the Royal Dutch Water Network (KNW) having 4000 members, which is substantive knowing that the total number of IWA individual members is 10.000. In addition, the Royal Institute of Engineers (KIVI-NIRIA) has approximately 2100 members with a background in water management and water engineering.

The country spent 1.6% of the GDP on tertiary education compared to 0.8 in the United Kingdom in 2009 and 1.2 in the United States in 2009 (UNESCO Institute for Statistics, 2012) The lowest value is 0.8 and the highest is 2.4 for North-West Europe and the United States, placing the Dutch expenditure at average. For Indonesia this figure was 0.4% in 2009. In South East Asia and the Pacific, 0.3 was the lowest value for 2009 and 2.2 the highest (UNESCO Institute for Statistics, 2012). Hence, Indonesia tends to under-invest in its intellectual capital. The number of specialised educational programs leading to a degree dedicated to a water related field in the Netherlands is relatively limited, but many programs exist that offer a water specialisation (Adviescommissie Water, 2012). At the institutional level, governance systems allow stakeholders in society to exert pressure on the government to pay attention to certain issues that may not have been on the agenda previously. This civil society pressure can serve as a KCD mechanism as well and I will give examples of this when I describe different phases in Dutch water management history (see Section 7.3).

The joint report by the Dutch Council for Agricultural Research (NRLO), the Advice Committee for Science and Technology Policy (AWT) and the Council for Environmental and Nature Research (RMNO) on knowledge and innovation for water management in the Netherlands (de Wilt et al., 2000) has described the water sector comprehensively for the field of integrated water management. Though this document was published in the year 2000, the report still offers a good indication of the knowledge institutes and KCD mechanisms available in the Dutch water sector.

The report identifies 35 water knowledge centres, consisting of 8 national institutes, the water boards (currently 25), 5 universities, 3 para-academic institutes, 2 Netherlands institutes for applied scientific research (TNO) and 1 large technological institute (GTI), 8 companies delivering advisory services, of which 2 conduct research, and 6 are consultancies. Together, these centres are responsible for more than 60% of the knowledge development, represented in 800 projects. The National Institute for Freshwater Management (RIZA) and the National Institute for Coast and Sea (RIKZ) have merged into the Water Agency, under the Ministry of Infrastructure and Environment. A small part of these institutes have per 1 January 2008 merged with Delft Hydraulics, GeoDelft and TNO-NITG (TNO-Central Geoscientific Information and Research Institute) to form Deltares. The Dutch water sector, covering the fields of water and river management, drinking water, water and wastewater treatment and infrastructure development, employs around 80.000 people (2012). Dutch delta technology provides a yearly turnover of 7.5 billion euro in 2008 (Muizer et al., 2010), and Dutch water technology delivered a turnover of 9.1 billion euro already in 2003 (Muizer and Leusink, 2005). For this dissertation I focus on subsectors of the water sector that focus on water quantity, such as river management, flood protection, and coastal management, and I exclude fields such as drinking water and (waste) water treatment. I further focus on the national level only. For my organisational analysis I investigate the Rijkswaterstaat specifically. With the Rijkswaterstaat I mean the implementing organisation of the Ministry of Infrastructure and Environment, but also the Rijkswaterstaat as it functioned before 2002, when it was not only an implementing organisation but responsible for policy formulation as well.

Based on Kuhn's seminal work on shifting scientific paradigms (1962) (see also Chapter 4, Section 4.2.1.), and further work on transitions and policy change (de Haan and Rotmans, 2011; Van der Brugge, 2009; Huitema and Meijerink, 2009b; Kingdon, 1995) a number of distinct development phases can be discerned in Dutch water management. In the Dutch water sector a few significant 'revolutions' have taken place in which the practice of water management changed profoundly, heralding the replacement of the prior paradigm with a new one. These were driven as much by shifts in the social-political as in the scientific-technical spheres. I use theory on policy change (Kingdon, 1995; Sabatier and Jenkins-Smith, 1993) to further explain how these paradigm shifts in policy come about and distinguish three phases that form the background against which organisational change occurred and the functioning and mechanisms of KCD in the Dutch water sector that are discussed, can be understood better.

7.2   RESEARCH STRATEGY AND METHOD

7.2.1   Case selection

The Dutch case study was selected to serve as a reference case for the Indonesian case study. The analysis does not pretend to exhaustively analyse the knowledge and capacity requirement of the current the Rijkswaterstaat and/or the sector priorities or make pronouncements on whether changes in the sector and in the organisation have been effective and appropriate. More so, this analysis is used to explore the boundaries of the validity of the adapted KCD conceptual framework that we have derived to study KCD in developing countries and to generate a reference to which to compare the results of the Indonesian case.

7.2.2 Research strategy

For an elaborate description of the research strategy for this case study, I refer to Chapter 3.

In this case study I apply the same theoretical framework that I use in the Indonesian case. Because this case study serves as a reference case for the Indonesian case study and because the Dutch sector has been studied in some depth by others, this case is not analysed as comprehensively. For the institutional level I adopt the Multiple Streams Framework (MSF) by Kingdon (1995) and the Advocacy Coalition Framework (ACF) by Sabatier (1993), described elaborately in Chapter 4. The institutional longitudinal analysis is based on literature, partly on semi-structured interviews and specifically for the section on stakeholder involvement, on the quantitative survey.

At the level of the organisation, I take up theory on organic and mechanistic organisational structure by Burns and Stalker (1961), described in Chapter 5. To describe the organisational structure I based my investigation mostly on literature and for the description of KCD mechanisms at the organisational level, I used the survey and interview results.

Finally, at the level of the individual I adopt theory on the concept of competence (Cheetham and Chivers, 1996) and on T-shaped competence profiles (Oskam, 2009), explained in Chapter 6. My analysis included literature about the type of knowledge that is appreciated in the Rijkswaterstaat and the mechanisms that individuals use to acquire different types of knowledge and capacity. Although I provide an indication of the level of aggregate competences and meta-competence for continuous learning and innovation for the organisation as a whole, I do not provide a detailed analysis of these competences at the individual level, nor of the subcomponents of competence.

7.2.3 Method

7.2.3.1 General

The methods I used for this case study are described in Chapter 3, Section 3.3 and are part of a so-called mixed method approach, consisting of qualitative data collection through semi-structured interviews, desk study of literature, and quantitative data collection by means of an online questionnaire. I conducted 6 interviews with experts of the Netherlands water sector that were consequently analysed with Atlas.TI as explained in Chapter 3 (Annex C). As mentioned, a disadvantage to online questionnaires is the often low response rates. Although much effort was made to advertise the questionnaire, I experienced that here as well and therefore not all results are statistically significant at a 10% significance level. To analyse the extent to which variation can be attributed to real differences instead of random fluctuations in the sample, tests of difference and analyses of variance (ANOVA) were performed. Differences between cohorts that are statistically significant at a confidence level of 10% are indicated with an asterisk. Where no asterisk appears it cannot be excluded statistically that observed differences are attributable to chance; however, they can indicate patterns. In combination with the qualitative data, the complete dataset provides a basis for analysis.

7.2.3.2 The online survey

The set-up and strategy for the survey is explained in Chapter 3, Section 3.3.4 and 3.3.6. The online questionnaire (Annex D) generated 155 responses. Of the 155 responses, 120 were male (77%) and 35 female (23%). The largest percentage of women falls in the age group of 35 years old and younger. Forty percent of this age group is female. The age group older than 55 includes one woman. The percentage of women in the Netherlands that chooses a technically oriented (technology, industry or construction) post-graduate education and graduates, is 24,4% according to the Central Bureau for Statistics (Tijssen et al., 2010) and although these two groups cannot be compared one-on-one, the percentage in my sample is representative for the population. The youngest respondent in the sample was 25 years old and the oldest was 70 years old. Most of the respondents fall in the category 45 to 54 years old. In the water sector as a whole in the Netherlands, the average age may be slightly higher but the sample seems representative enough also in this count. Of the respondents of this questionnaire 65% were coming from the public sector and 30% from the private sector. For the remaining 5% this was unknown.

Forty percent of the respondents is working in policy making, 34% is in a management position, 32% in research and 20% and 16% in support functions and in implementation of projects and programmes respectively.

When dividing the sample group over subsectors, the largest percentage (38%) is working in integrated water management, 26% in drinking and wastewater, 22% in water quality management and the remainder of the sample in agricultural water management, water quantity management such as flood management, hydrology and climate change adaption, and construction of water-related infrastructure. If the sample size would have been larger, statistically significant results could possibly have been derived per subsector or function. As this is not the case the sample group is treated as a homogenous group.

7.3 THE INSTITUTIONAL DECOR: SHIFTS IN DUTCH WATER MANAGEMENT

7.3.1 The territory

The Netherlands is a small, densely populated country with an area of 37,400 sq km and a population of 16.7 million people, per 1 January 2012. The country lies in the delta of three major North-West European rivers: the Rhine, the Meuse and the Scheldt. Annual rainfall is about 750 – 800 mm, with a deficit in the summer months. The largest population concentrations are in the low-lying urbanised areas in the west of the country along the coast, including cities such as Amsterdam, Rotterdam, The Hague and Utrecht. The Netherlands has a long water management tradition, in land reclamation, flood protection, and managing the quality and quantity of the ubiquitous water bodies. The history of water management in the Netherlands has been described elaborately (Lintsen, 2002; Van de Ven, 2004; Bosch and Van der Ham, 1998; Huisman, 2004; Van der Ham, 1999; Toussaint, 2003; Rooijendijk, 2009). I reviewed the events that are relevant for the discussion of the development of knowledge in the sector, and use this as framework to analyse the results from the interviews and literature that provides insight in institutional development and policy change.

7.3.2 Three phases

Many authors have distinguished phases that help to emphasise the major changes that have taken place over time in Dutch water management. As described earlier, I apply Kingdon's MSF to indicate paradigm shifts in policy making. Kingdon, in line with Kuhn's paradigm theory, argues that a stable situation will last until other coalitions have found a way to put different issues on the agenda. The various coalitions then represent the struggle amongst different kinds of knowledge, and knowledge bearers for dominance. What knowledge is relevant in the new situation, and who is entitled to be an authority (Disco, 2002)?

Several authors have studied the development post-1950, some from a transitions perspective (van der Brugge et al., 2005; van Heezik, 2008; Blankesteijn, 2011; Lintsen et al., 2004), others from professional experience (Huisman, 2004; Bosch and Van der Ham, 1998). A number of interviews[39] indicate that beginning of the 1970s is seen as associated with a significant shift, from a very mono-disciplinary technocratic period to a more multi-disciplinary period that included environmental values in water management. This is described in the literature (van der Brugge et al., 2005; van Heezik, 2008; Blankesteijn, 2011; Lintsen et al., 2004; Bosch and Van der Ham, 1998). For the period of the 1990s and onwards, depending on the focus and scale of the research, researchers define paradigms differently. Heezik (2008) takes the floods of 1993 and 1995 as a turn in water management, as these highlight river management. De Wilt et al. (2000) mention the turn of the century as a policy and mental shift from protecting the Netherlands from water, to accommodating the water. However, their arguments for this shift are not based on a theoretical explanation. Using the MSF, I define the period from 2002 until present as a new paradigmatic phase, because politics put pressure on the Rijkswaterstaat (the Rijkswaterstaat as implementing agency of the Ministry of Transport and Public Works, see Section 7.3.5) to make rigorous changes in its organisational management and policy, to become more accountable and efficient. In the next sections, I describe the water history preceding the first phase and then continue in each section by explaining how changes in policy mark the beginning and end of the first two phases. I explain how new knowledge and capacities in turn are used in coalitions to pursue policy change and lead to a new phase. The third phase is less the result of a change in policy than it is a radical organisational change under pressure of politics, driven by the pursuit of more accountability and efficiency in the public sector, a movement that started in the 1990s already, leading to a substantial shift in the mix of knowledge and capacities.

7.3.3 Dutch water history – before 1950

At the beginning of the nineteenth century Dutch water management practices were oriented to flood protection and reclaiming land. Between 1833 and 1911, 3500 km2 of land were brought into cultivation, of which 1000 km2 had been reclaimed from the sea and drained lakes (Van de Ven, 2004). Technological developments in the early twentieth century allowed the large Zuiderzee (the Southern Sea, an inland sea in the North-West of the country) project to be closed by constructing a dike of 32 km, which would lead to the protection of the hinterland from coastal flooding opening opportunities for land reclamation and land development. The flooding of 1916 around the Zuiderzee and the manifest vulnerability of the Dutch food supply during the First World War from 1914 –

---

[39] PD70, 71, 74

1918, accelerated the decision making. The Law of 1918 to close off and reclaim parts of the Zuiderzee, not only resulted in a new approach to coastal defence, but also represented a first step towards a truly national policy in the field of water management. The dike was finished in 1932.

7.3.4   Phase I: 1950 -1970: 'Development of water infrastructure'

Older engineers look back with melancholy to the period prior to 1970. It was a period of national effort and public esteem for them (Lintsen et al., 2004; Blankesteijn, 2011), recognised in the country as well as abroad. The esteem was based on the success stories of reconstruction and rehabilitation of war damage after the Second World War, the construction of the Delta Works after the tragic storm surge in the South-West of the Netherlands in 1953, and new transport infrastructure that was constructed to keep pace with and facilitate increasing prosperity in the country. The impressive Delta project, and the way in which the Rijkswaterstaat handled the storm surge and resultant flooding of 1953 contributed greatly to its prestige (Table 7.1). The organisation grew in size as a result of all the construction works and gained significant experience in innovative flood protection techniques.

The long-standing efforts to control water brought safety and prosperity, and the successes in achieving this goal in the period between 1953 and the new century instilled a measure of faith in the technologies and institutions used to accomplish this (Huitema and Meijerink, 2009a). Large scale innovations were implemented in this period in coastal engineering, such as the first artificial Meuse plain for expansion of the Rotterdam port. The civil servants of the Rijkswaterstaat grew very powerful in this period and accustomed to getting things their way. They would tend to provide information about their plans to the outside world often at a very late stage, leaving little room for negotiation or adaptation. Policies were determined by the bureaucrats in the Rijkswaterstaat, rather than by politicians (Van den Brink, 2009), although Blankesteijn (2011) mentions that in water management in those days, the politician could 'embrace' the technocrat because their work was in his interest too. In this context, knowledge served a direct economic goal, namely the development of the agricultural and industrial sectors. This is similar to the Indonesian case study, where the first phase that we identified after the Second World War is likewise characterised by infrastructure development to serve economic development (Section 4.3.1).

There was a very strong esprit de corps and achievements were seen as the results of teamwork. In his insightful book, Metze (2008) describes how many of the chief Engineers-Directors had been acquainted from their university days, and how they looked out for each other. New Directors-General always came from the same circles and made sure that politics were kept at a sufficient distance. In the current Rijkswaterstaat The seventeen Chief Engineers-Directors are perceived as regional 'barons' with a significant financial budget (Metze, 2008).

The position of the engineer changed gradually between 1950 and 1970. Whereas before the middle of the twentieth century an engineer was responsible individually for a design or work, the individual had now been integrated into a team of specialists. The Construction Services of the Rijkswaterstaat made the design and evaluated it; their staff co-operated with specialists from institutes like the Delft Geotechnics Laboratory, Delft Hydraulics Laboratory or the government Institute for Drinking–Water Supply because their expertise was essential. A need for integration, and thus teamwork, in this period in mostly technical subjects, started to arise. Most of the employees were engaged in planning and doing the construction, maintenance and the management of infrastructural and hydraulic works.

The fast technological developments and welfare led to better accessibility of education and the level of education of the population increased towards the end of the 1960s (de Heer, 1991). The population, better able to express its ideas, and more critical, was by that time increasingly worried about water pollution and environmental health. Their perception of the Rijkswaterstaat started changing. Even though the organisation was already beginning to become more 'green' and ecological awareness was growing, the Rijkswaterstaat was seen as a state within a state (Bosch and Van der Ham, 1998; Lintsen et al., 2004; Van de Ven, 2004; Van den Brink, 2009). It was seen as a technocratic apparatus that wanted to keep control and had its own strong agenda. The organisation was also granted this space by politicians. The changing perceptions increased the pressure on the traditional Rijkswaterstaat professionals.

7.3.5 Phase II: 1970 –2002: Towards IWRM and better governance

7.3.5.1 Pressure from society

In the beginning of the seventies the functioning of the government and 'the establishment' were increasingly criticized as the US and Europe were engulfed in a mild cultural revolution. Old paradigms were openly questioned, the voice of the citizen promoted and the technocratic assumptions of development doubted. Therefore, the policy of economising was combined with a reduction of the role of government and the reinforcement of the private sector. The self-confidence the Rijkswaterstaat decreased and people maligned the Rijkswaterstaat as an organisation pushing its plans regardless of the effects[40] (Table 7.1). Literature shows consensus on the external pressures that forced the Rijkswaterstaat to make a change towards incorporating environmental values (Huisman, 2004; Lintsen et al., 2004; Bosch and Van der Ham, 1998; Metze, 2008; de Heer, 1991; Disco, 2002). The socio-political climate in the Netherlands changed; politics were in motion, new political parties were established and political parties had to renew under pressure of the 'young guns'. Political parties embraced environmentalism, as nature and environment became more prominent in society. This is illustrated by the large scale protests against the closure of the Eastern Scheldt and the construction of the motorway A27 through the Amelisweerd estate in 1978 (Lintsen et al., 2004). Whereas it was first praised as a major triumph of civil engineering, the Eastern Scheldt storm surge barrier suddenly came to be defined as an environmental catastrophe (Disco, 2002). The Rijkswaterstaat found itself in a downward spiral of criticism, and reduced manpower, which was reinforced by the ailing economy that didn't improve until 1985.

An important accelerator of KCD that makes its appearance for the first time here is the pressure of well-informed and vocal activist groups that push their agendas, in this case environmental issues. They acted as 'the conscience of society'[41]. They proved to be a powerful coalition (Sabatier, 2007) that managed to put their concerns on the political agenda, and so ensure that knowledge was mobilised to arrive at solutions. In line with

---

[40] PD71
[41] PD71

the MSF, we see a political stream and a problem stream coming together, forcing the policy stream to change.

7.3.5.2 Innovation in the Delta

After public debates in 1974 and the resulting compromise of a flexible storm surge barrier, the Rijkswaterstaat demonstrated a reorientation in the sector of coasts and estuaries with the introduction of environmental values in its design of the Delta Works. For rivers and roads, this paradigm shift took much longer. In the coasts and estuaries sector, the Delta Agency of the Rijkswaterstaat that had been established by the end of 1969 appeared as an important forerunner. Whereas in rivers and roads changes were limited, in the coasts and estuaries sector a whole new regime developed. This occurred as part of an exogenously enforced internal learning process, with protected space for try-outs in the Delta Agency. The Delta Agency inventoried all issues related to the design of the southern part of the Delta Works. For the Eastern Scheldt storm surge barrier specifically a number of project groups existed in the Delta Agency. They worked on the management of the works, the design, engineering, multidisciplinary groups and environment groups. Discussions were common between the project groups that served to use the tacit knowledge available in the project groups and to collaboratively find solutions for problems. This process has much in common with social learning, although not consciously named so. Increasingly, the water sector was in need of systems thinking, because water management had become increasingly complex after the 1960s. New facets were included in water management, such as environmental knowledge, transboundary management with other European countries, and biodiversity. Furthermore, stakeholders increasingly wanted to be informed and influence decisions. The first steps in this way of thinking were made in these project groups. The groups were coordinated by a steering committee that consisted of Chief Engineers-Directors and other leaders of involved institutes.

If the questions were very complicated, other institutes were engaged to conduct further research. Most of the fundamental research was outsourced to knowledge institutes such as the Delta Institute for Hydrobiological Research in Yerseke (now the Royal Netherlands Institute for Sea Research – Yerseke), although there were also applied research groups in the Delta Agency.

The Rijkswaterstaat hired Rand Corporation, an American think tank for strategic management, to generate new knowledge in the field of coasts and estuaries, and to supplement their lack of knowledge of ecological systems, policy analysis models and the use of computers for modelling of for example water distribution and sediment transport (Rand Corporation et al., 1981). The majority of interviewees mention that Rand Corporation was the first to introduce systems thinking for decision making in complex environments. 'Rand Corporation helped us to think about alternative solutions'[42].

In the Delta agency, staff were developed through training, but also through coaching and the manner of leadership. Rand Corporation also advised on management and staff development approaches. The approach was personalised to the needs of specific staff members[43]. The Delta agency took the role of incubator, in which time and resources

---

[42] PD70, PD71, PD73, PD74
[43] PD71

were available to learn and reflect. This was arguably the first time in the history of the Rijkswaterstaat that learning was so consciously arranged.

7.3.5.3 Space for other disciplines

The increasing attention for the environment resulted in a change in tasks. The integrated water management that was propagated and the emphasis on water quality resulted in an influx of ecologists in the Rijkswaterstaat (Van de Ven, 2004), bringing along other networks and different attitudes.

Whereas in the first phase a high appreciation for substantive technical knowledge and experience in a narrow definition of infrastructure development was revealed, from the 1970s onwards other fields need to be taken into consideration, representing technical substantive knowledge on amongst others ecology and the environment, spatial planning and eventually organisational management. Social sciences only gained importance slowly, mostly because water was still considered a strictly technical matter, with a strong focus on safety aspects and development. Furthermore, stakeholder involvement, and awareness of the need to take into account societal developments, require knowledge of different aspects of social science, and were not common practice yet. In the 1980s the new aspects of water management started to take form in the Second and Third National Water Policy in IWRM (Ministerie van Verkeer en Waterstaat, 1986). This policy was a new vision in which rivers, lakes, seas and all other waters are treated as connected ecosystems. The preparation of the Third National Water Policy was accelerated because the Minister, Mrs. Smit-Kroes, wanted it completed before she stepped down from her function as Minister of Public Works, of which the Rijkswaterstaat was part. This provided the political leverage needed to enable and facilitate the policy process, allowing policy entrepreneurs to make the necessary coupling between the political and problem streams and place issues onto the policy agenda.

With the Third National Water Policy, the new knowledge of the 1980s got consolidated in a new policy regime of IWRM, for coasts and estuaries and for rivers as well. Van Heezik (2008) describes how the rivers sector experienced more difficulties because on the one hand there was an established coalition of Water Boards, provincial departments of public works, and the Rijkswaterstaat, focused on the classical approach of dike strengthening and representing the status quo. On the other hand, a coalition consisting of action groups, trusts and NGO's formed, brought in new substantive knowledge on nature conservation, and was able to more effectively mobilise public opinion, to focus on conservation of landscapes and monuments and arguing against large dike and river works that disrupted landscapes and changed ecologically vibrant river systems in concrete canals. The technical and institutional issues proved more complex than those in coasts and estuaries. The two different coalitions barely communicated with each other, resulting in a stand-off. Two government committees were installed as intermediaries, but without substantial success.

7.3.5.4 Stakeholder involvement

Stakeholder involvement had become more important over time in Dutch water management, at least in public discourse, starting with the 'Ecological Turn' described by Disco (2002). It was nevertheless still rare in the river subsector in the 1990s. In the river subsector the technical paradigm still prevailed, with neglect of other interests that were represented by the other coalition.

The interviews, and from 1990 onwards the sector-wide survey, clearly reflect the increasing influence of stakeholders sector-wide. Figure 7.1 shows four different intensities of stakeholder involvement.

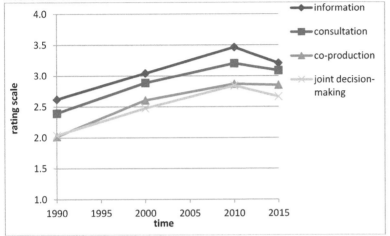

Figure 7.1. Involvement of stakeholders as rated by all respondents on a scale from 1 = not at all, 3 = to some extent 5 = extensively)

These intensities are defined based on the participation model initially designed by Arnstein (1969), and here adapted from Denters et al (2003):

- Information: authorities inform stakeholders about their policies and decisions, (lowest intensity)
- Consultation and advice: opinions and preferences of stakeholders are gathered and taken into account in the process of decision making, and can be binding or not, (medium intensity)
- Co-production/co-creation: stakeholders are invited to participate in the process of policy design, (medium-high intensity)
- Joint decision-making: Formal or informal arrangements have been designed that give stakeholders the position of (co-)decision-maker (highest intensity).

Although the differences between the scores and intensity of stakeholder involvement are not significant at a 90% confidence level, the data indicates a trend that is confirmed in half of the interviews[44].

The modest scores indicate that although the role of stakeholders has become more important over the last two decades, and the public discourse may give the impression that public participation is common practice in water management in the Netherlands, the actual level of participation is perceived to be very high, albeit growing steadily over the phases. Especially 'co-production' and 'involvement in decision making' do not happen regularly, and are still not common in Phase III. Especially consultation and-coproduction would lead to the inclusion of a different type of knowledge, and could therefore be valuable. The Fourth Policy on Water Management (Ministerie van Verkeer en Waterstaat, 1998) recognises that involvement of citizens is important, but the policy

---

[44] PD70, PD71, PD74

mostly stresses information provision, and not the other forms of participation. In the policy document prepared by the Committee 'Water in the 21st Century' (Directoraat-Generaal Rijkswaterstaat, 2000), stakeholder involvement is more prominent: Not only informing stakeholders, but also consulting them in the planning phase of flood mitigation and prevention measures. In the context of the EU Water Framework Directive, active participation, extending beyond informing stakeholders, reportedly was not realised in 2006, according to the European Environmental Bureau (2008). It thus appears that the low figures for stakeholder involvement are confirmed by the policy documents on Dutch and European Water management. It is unclear why the prognosis for 2015 is unanimously scored lower than for 2010. Perhaps some pessimism is creeping in, because co-producing knowledge with different stakeholders has shown to be difficult to realise and is very complex (Edelenbos et al., 2011), or perhaps a re-valuation of technical substantive knowledge is predicted to start overruling arguments for increased stakeholder participation.

7.3.5.5 Closer to IWRM

The earlier mentioned report by the NRLO, the AWT and the RMNO gave advice about the knowledge and innovation requirements for good water management in the Netherlands (de Wilt et al., 2000). The report indicates the need for professionals with a broad vision, who combine thorough specialist knowledge with affinity for diverse 'cultures', i.e. cultures of different stakeholders and disciplines. At the end of the second phase professionals were still driven by a relatively technical-scientific approach to water management and knowledge on the social aspects of water management was not widely available. The cultural change that De Wilt et al. encourage for the new century involves a change from the Rijkswaterstaat as a relatively closed stronghold, focused on construction and development, to an organisation to work with a paradigm based on IWRM. (de Wilt et al., 2000). Later in this Chapter I will discuss this in the context of individual competences.

In addition, in the mid-1990s tensions emerged between water management and spatial planning. At the Ministry of Housing, Spatial Planning and Environment, the spatial planning department was reluctant to accept that water managers were going to take into account the space needed for water, as described also by one of the interviewees[45]. The floods of the Rhine and Meuse rivers and near-disasters of 1993 and 1995 instigated a radical turn in river management: the Delta Plan for Large Rivers, in analogy to the 1953 Delta Plan to protect the coasts (Lintsen et al., 2004). In this plan already a more prominent role was given to the opinion of stakeholders other than the technical professionals.

In spatial planning, changes took longer. In 1997 an attempt was made to change article ten of the Spatial Planning Law, which specified that spatial planners needed to involve the water manager, 'when necessary'. The water management department in the Rijkswaterstaat wanted to remove these two words, to make sure there was mandatory cooperation between spatial planning and water management. This turned out to be very difficult, until 1998, when heavy rain created a new threat of floods and politicians decided that cooperation between spatial planning and water management was imperative. This also meant an increased role for stakeholders, such as nature

---

[45] PD74

conservation groups, and local citizens living in the areas under threat. After this important decision, the development in effect led towards more space and expensive land being set aside for water, resulting in the policy response 'Room for the River'. This was further consolidated in the Fourth Policy on Water Management (Ministerie van Verkeer en Waterstaat, 1998).

To summarise, the changes that occurred in Dutch water management roughly until the turn of the century, concern primarily the broadening of the technical scope of work by pulling in new technical disciplines to help address the increasingly complex problems involving the environment. The Rijkswaterstaat had to accept the involvement of more disciplines to come to better results, leading to a better integration of water and nature, i.e. the 'Ecological Turn', of water in relation to its chemical and biological quality (Surface Water Pollution Act); and of land use and climate change adaptation, resulting in 'Room for the River'. Secondly the inclusion of a multitude of different stakeholders also forced the Rijkswaterstaat to develop the competence to work with them.

7.3.6   Phase III: 2002 – present: Accountability and efficiency

7.3.6.1   *Wider societal and political drivers*

This phase is driven by a second type of change, to a more accountable and efficient management of the sector. The Washington Consensus, a term to indicate policy reforms instigated by institutes mostly based in Washington, such as the International Monetary Fund (IMF), the World Bank and the American Treasury, and in the UK started to advocate more democratic control and accountability, better budget discipline, and a larger role for competition and the market. The Consensus argued for 'small government' and deregulation of the economically productive sectors. Later the Washington Consensus got a more negative connotation over emphasising a neoliberal agenda.

The trends set in motion by the Washington Consensus were amplified by other think tanks such as the OECD and by the Davos Economic Forum and had a significant influence on Dutch policies well. Many national services were privatised, such as the national airport, the postal services and KLM, while others were deregulated such as the national railways, and several social services.

In this political climate, two major broad perceptions had emerged during the 1990s: the close contact between the Rijkswaterstaat and political circles and society seemed to have loosened and the Rijkswaterstaat was appearing less sensitive to societal tendencies. Furthermore, following several fraud cases in the construction sector and highly publicized excessive overruns of budgets for the new high-speed railway and for the cargo railway through the Betuwe, doubts were raised in political circles about the expertise and reliability of the Rijkswaterstaat (Metze, 2008).

The societal and political drivers that finally led to changing fortunes for the Rijkswaterstaat were the drastic political shifts partly as a result of Pim Fortuyn's influence on politics. Fortuyn channelled the dissatisfaction of the Netherlands electorate with the current government, seen as distant from the public and elitist. He criticised the public sector heavily, railing against their struggle with crime and safety, the lack of quality in education, traffic jams, and so on (Fortuyn, 2002; Otjes, 2011). 'Against the decor of terror attacks and disasters in 2001 and 2002, he proclaimed that Dutch society deteriorated in the 1990s as a result of slack policy' (Metze, 2008; Fortuyn, 2002).

He was assassinated on 6 May 2002, and his ghost hovered over the Rijkswaterstaat. Two of his complaints about the public sector fell on fertile ground: first that the government had lost contact with the people, and secondly, that they needed to outsource many more tasks to the private sector, as they had to work more efficiently (Metze, 2008). Where the neoliberal policies of the Dutch governments harked back to the Washington Consensus, a shift for the Rijkswaterstaat required a final trigger that came in the form of Pim Fortuyn's populist pamphlets.

This period can be considered instrumental for the political stream and the problem stream to come together and create a window of opportunity for a new management regime (Kingdon, 1995) and the Rijkswaterstaat was subjected to reorganisation.

7.3.6.2 A new organisational direction

In January 2002 a new organisational structure was formalised. The two most important functions of the former the Rijkswaterstaat, policy preparation and implementation, were separated and located in different institutions: the Directorate-General Water, the Directorate-General of Public Works and Water Management, still called the Rijkswaterstaat, as well as the Inspectorate (Van de Ven, 2004). The new Directorate-General Water was to develop water management policy and regulation whereas the Rijkswaterstaat would keep all executive tasks. In 2003, a large scale internal reorganisation of the new Rijkswaterstaat was launched, that lasted until 2008.

This type of change was very different from the incorporation of ecological values and spatial planning in water management and the introduction of 'Room for the River' which can both be viewed as an expansion of substantive knowledge to support a more integrated natural and physical systems approach (Table 7.1). In the new paradigmatic shifts and deep re-organisation, the key objectives were efficiency and accountability. Improved efficiency was supposedly gained through the establishment of Public-Private-Partnerships, privatisation, and better cost recovery, and budget control and reduction of overhead costs from about 31% to near 20%. The staff would be cut by about 3000 members over 4 years. Accountability was pursued by increased control and audits, more attention towards 'serving the citizen' (through call-centres, websites were opened, and repair of damaged roads done faster) and increased participation of stakeholders. As a consequence, the knowledge demand emphasised knowledge about organisational management, process management, providing frameworks and connections for stakeholder networks, and becoming a professional outsourcer (Van de Port and Veenswijk, 2006). Table 7.2 gives an indication of how the organisation was supposed to change according to the newly appointed Director General of the Rijkswaterstaat, Mr. Keijts. Although some of the items are quite vague, e.g. 'constraining organisational management' vs. 'organisational management that helps us', it reflects the major cultural change that was forced upon the Rijkswaterstaat.

Until the reorganisation of 2002 - 2008, the Rijkswaterstaat functioned largely in the same way as before. Over the years the knowledge base had significantly expanded, and with the introduction of internet, communication became easier. The working culture did not change in that period. Metze (2008), in his book 'Changing tides', a well-documented account of the reorganisation, concludes that it was felt that the organisation was too bulky, too expensive and not sufficiently transparent. Along the same line, several interviews[46] indicate that before the reorganisation, staff had substantial autonomy. It was clear who was responsible for what and staff knew that they could talk with other ministries without consulting their boss first.

The reorganisation changed this culture and way of working to a more control-oriented situation. In the Rijkswaterstaat it had not been common to openly criticise one another, nor to expand on mistakes. This behaviour is to some extent comparable to the Indonesian case. Mr. Keijts wanted this to change. The Rijkswaterstaat culture valued loyalty, the old culture was that you assumed you understood each other and left it that way (Metze, 2008). In the new culture, Keijts wanted it to be: 'say honestly if you can do something, and say it much earlier.' In the organisation however, this was experienced as top-down control. Every plan, idea or activity in Phase III had to be endorsed by the management team of the Rijkswaterstaat. Whereas the accountability proved a good aspiration, the new intensive scrutiny at central level led to authority lines becoming much longer, according to interviewees[47]. The management team was often perceived as quite harsh and overly critical to the employees. Taken together, according to interviewees this led to a situation where people did not want to stick their neck out.

Table 7.1. The proposed attitude and skills shifts during the reorganisation of 2002 of the Rijkswaterstaat (Rijkswaterstaat, 2004)

Before	After
Content oriented manager	Public oriented network manager
Focus on internal needs	Focus on user
'King' of the province	Team player, equitable partner
Reinventing the wheel	Coherence and cooperation
Cooperating superintendent/boss	Professional outsourcer/buyer (contract engineering
Supply-driven knowledge	Demand-driven knowledge attached to main products
All knowledge in-house	Smart organisation knowledge (externally)
Constraining organisational management	Organisational management that helps us
Avoiding (HR) problems	Tackle (HR) Problems

Interestingly, this led to the opposite of what Mr. Keijts wanted to achieve, as people could not respond honestly regarding their capability to complete assignments or not. The ensuing rapid downsizing of the organisation and the recruitment of new staff with the new process management skills in effect led to a loss of technical substantive

---

[46] PD71, PD74
[47] PD71, PD74

knowledge. At any rate, more process managers were appointed in decision-making positions so we may conclude that technical substantive knowledge was receiving a less prominent role in decision making.

At the beginning of the change process the management felt that the majority of the staff had grown too much into a conventional technical mould and did not manage to adjust to the new education and quality demands of the organisation. To meet the quality demands, the Rijkswaterstaat temporarily, but in practice for long periods, hired professionals from outside. The new policy was to shrink the organisation by 30% through early retirements, no renewal of temporary contracts and no external hiring. In theory, this would provide space for four hundred to twelve hundred new employees with the appropriate educational background and experience (Metze, 2008).

Many senior professionals opted for early retirement, leading to an exodus of tacit knowledge. It took considerable time before new professionals were hired. However, the newly hired professionals had a different set of competences, and although valuable, as such it did not replace all the knowledge lost with the departure of the senior staff.

There are also some annotations at the top (rotated): "...of 1953", "activists, political shift", "2003 - 2008".

Development of the Rijkswaterstaat	Phase I: Development of infrastructure	Phase II: Towards IWRM and better governance	Phase III: accountability and transparency
Time line	1950 — 1970	1980 — 1990 — 2000	2010
Institutions:			
National context and political dynamics	Rebuilding the Netherlands, high work ethos, infrastructure for increasing prosperity	New young generation without war experience, time for recreation, increased role of civil society, TV becomes important medium for communication, central left wing government, influenced by environmental lobby, introduction of internet, increasing pressure on space, climate change, Washington consensus: decentralisation, privatisation	Increasing use of internet for knowledge exchange, e.g. social networking and media, CoPs etc., Large scale infrastructure project failures, construction fraud, right wing government, transparent and efficient government, critical view on government efficiency
Water sector dynamics	Rebuilding the Netherlands Flood control	Focus on integration of disciplines, resulting in IWRM, preparedness for climate change, stakeholder involvement, new approach to river management	Large scale infrastructure project failures, construction fraud, leading to increased outsourcing for greater efficiency. Becoming a 'public oriented network manager'
Culture in the Rijkswaterstaat	Strong esprit de corps, technocratic, state within a state, hierarchic, little collaboration and communication with non-engineering actors	Proud self-image was broken in the beginning of the 1970s, reflective capacity, space for learning, increasing collaboration with non-engineering fields, increased communication, remains of the old Rijkswaterstaat culture until reorganisation	Increased attention for knowledge management in a later stage of reorganisation
Organisation:			
Main legislation and Policies	1st National water policy (1968): increasing water supply with large scale infrastructure development	2nd national water policy (1985): economic approach to water supply + water quality 3rd national water policy (1989): water quality and quantity, nature protection, European Water Framework Directive (2000), 4th national water policy (1998) based on IWRM	Water act (2009), preparation of Environment Act (2013?), the National Water Plan (2009)
Innovations	Coastal engineering mega-structures	State of the art structures taking into account ecological values, wastewater treatment, 'Room for the river'	'building with nature', tunnel technology
Individual level:			
Important competences	Civil engineering, hydrology, focus on flood control, safety, construction, focus on internal technical knowledge	Ongoing high expertise in engineering science, ecology, biology, systems approach to water management, introduction of IWRM, attention for external interdisciplinary knowledge, stakeholder involvement	Water governance, process-management, management of outsourcing, capacity to reflect on large scale infrastructure projects

Table 7.2. The Development of the Rijkswaterstaat in the Dutch societal context over a period of 60 years

127

7.4 THE INFLUENCE OF FORMAL ORGANISATIONAL STRUCTURE ON KCD IN THE ORGANISATION

7.4.1 The new order after 2002

In Chapter 5 I introduced a theory on formal organisational structure, and described the differences between an organic and a mechanistic formal structure and the influence on KCD. I will apply the same framework on this case, although in less depth than in the Indonesian case study.

Since the restructuring and partitioning of 2002, the new Rijkswaterstaat has become the independent executive arm of the Dutch Ministry of Transport and Water Management, anno 2012 renamed Ministry of Infrastructure and Environment. On behalf of the Minister and State Secretary, the Rijkswaterstaat is responsible for the design, construction, operation and maintenance of main infrastructure and policy support and technical advice. Ten regional departments (including 16 water districts) and one project organisation for 'Room for the River' (see Section 7.3.5) exist. Furthermore 5 Centres of Excellence exist of which one is the Centre for Water Management, the Waterdienst (Figure 7.2).

Together with the Deputy Director-General and the Chief Financial Officer (CFO), the Rijkswaterstaat's Director-General (DG) is responsible for managing the organisation. The DG is assisted in his duties by the Senior Management Advisory Unit. The senior management board consists of the executive board and the managing directors, also called the Chief Engineer-Directors (HIDs), see Figure 7.2 (Rijkswaterstaat, 2012).

7.4.2 Career system

Since the reorganisation, Human Resource Development (HRD) and Human Resource Management (HRM) for the Rijkswaterstaat are arranged by an internal dedicated department, a shared services centre and the influence that the Chief Engineer-Director exerted on HRD in the Phases I and II, was discontinued. As a result, contact between employees has become more distant and business-like. All services related to HRM, e.g. recruitment, career development and competence management, moved from the regional level to the central one.

Six types of functions are defined: line managers, advisors, project leaders, and implementing functions for water infrastructure and for roads, next to functions specifically for support and inspection. Staff in the first three functions can make promotion in a so-called triple ladder, meaning that an employee can make promotion in the type of function that he or she is in, and not necessarily only in line functions, creating a more organic structure with more flexibility for staff, and facilitating the flow of knowledge and experience. For exchange of experience and education of employees, a Corporate Learning Centre (CLC) and a Corporate Mobility Centre (CMC), located under the Rijkswaterstaat Centre for Corporate Services (Figure 7.2), were established (Rijkswaterstaat, 2010). The CLC develops education and training on specific subjects, or acquires courses from other parties. Training subjects include attitude, behaviour, skills and competences, and specific knowledge about cooperation with the private sector and other actors. According to the organisation itself, the CMC focuses on placing employees at suitable positions across departments and agencies (Rijkswaterstaat, 2010). The advantage of this measure is the enhanced mobility of knowledge across departments and regions, an indication of a more organic organisational structure. The downside is that, in practice people have little time to build up experience in a subject and to profit from that experience for a longer period. Career wise, they should not stay too long in one position; if they do, the assumption may be that they are not capable[48].

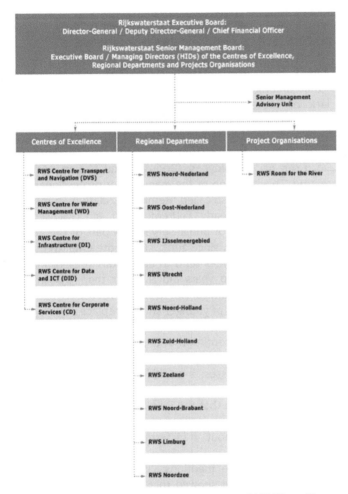

Figure 7.2. Organogram of the Rijkswaterstaat as per 2002 (Phase III)

7.4.3   Vertical and lateral communication

Through the three phases, the Rijkswaterstaat has increasingly felt the need to move to a more organic organisational structure, because the organisation was forced to communicate increasingly with actors beyond its borders. Communication has since the 1970s changed gradually from mostly instructions and decisions, i.e. vertical communication, to lateral, consisting of information and advice. As mentioned in Section

---

[48] PD74

7.3.6, before the reorganisation of Phase III, already many lateral connections existed between staff and between departments. As mentioned before, several interviewees report that communication with other organisations did not require prior approval from a superior. The interviews also indicated that the measures Mr. Keijts took to achieve a more open working atmosphere, instead also reinforced a control oriented situation, with characteristics of a mechanistic organisational structure, with a more hierarchic structure of authority and communication, and the location of knowledge of actualities exclusively at the top of the hierarchy as described in Chapter 5, Table 5.1.

7.4.4 Knowledge management in the water sector and the Rijkswaterstaat

7.4.4.1 The loss of tacit knowledge

Many interviews stressed[49] that among the most important perceived reasons for knowledge loss was the lack of succession planning both in respect of retiring professionals, especially in the third phase, and very high mobility in the Rijkswaterstaat in general.

Similar to what we found for the Indonesian case study in Phase III (see Section 5.3.3), many senior professionals, i.e. 1 out of 7 employees, (Boonstra and Muijen, 2011) will leave the organisation before 2016. This is on top of the many senior professionals that were lost in 2004 and 2005 as a result of the so-called 'Remkes-regulation', a measure that provided an attractive package for early retirement to all civil servants (InOverheid.nl, 2006) to decrease the number of public servants as part of the general reorganisation of the government. In any case, since 2005, 56% of the employees is 45 years or older, because since that year employees tend to again start their retirement later. The percentage of employees below 30 years old is lower than 5.5% in 2005 (Boonstra and Muijen, 2011). With the current political aim to slim down the bureaucracy, not much is done to bring in sufficient numbers of younger staff to replace seniors when they retire.

In the interviews, retired Rijkswaterstaat staff confirm that the reorganisation has led to a brain drain, or: 'Knowledge can be bought, but experience is something else' (Metze, 2008). The experienced water professionals possess a wealth of tacit knowledge that is very critical for the organisation. The Rijkswaterstaat acknowledges that: in Section 7.5 on individual knowledge and capacity, I explain that the Rijkswaterstaat wants to focus amongst others on maintaining or expanding the collective memory. The fact that so many professionals have since retired without transferring this knowledge is a breakdown of that collective memory.

All interview respondents[50] confirm this. One mentions: 'We undertook a strategic personnel analysis in the Rijkswaterstaat, and it showed that experiential knowledge is under pressure. This is knowledge of people working somewhere for a long time, who know the system. One of the reasons for the shortage of this type of knowledge is the belief in government organisations that staff needs to change to a different position every 5 years or so'[51].

---

[49] PD71, PD72, PD73, PD74
[50] PD70, PD71, PD72, PD73, PD74
[51] PD74

The lack of transfer to successors in general, is constantly high. For the Rijkswaterstaat specifically, the results are confirmed in literature as well (Metze, 2008; Boonstra and Muijen, 2011).

7.4.4.2 Capacity for outsourcing

In the Rijkswaterstaat, a crucial idea underpinning the reorganisation was the idea of 'the market, unless'. Unless it could not be done by the private sector, outsourcing would have to include not only the implementation of construction and maintenance, but also a large part of the preparation of such projects (Metze, 2008).

The interviews witness to the frustration felt by senior Rijkswaterstaat staff. Several interviewees[52] argue that the Rijkswaterstaat was manoeuvred into an impossible position during the reorganisation. They had to contract the assignments, and needed the knowledge and capacity to do so, but they could not have the human resources with the appropriate competences to manage outsourcing in a responsible way. The preparation and implementation of assignments was done by the private sector, but the Rijkswaterstaat was responsible for the quality. However, according to one respondent, the professionals who could judge the results had all retired or had been fired:

'I would always outsource certain assignments, but you have to do certain things yourself, otherwise you don't know what you are contracting. You cannot outsource properly if you haven't experienced that type of work yourself. For example, we had a laboratory, where all samples from national waters were analysed. At a certain moment it was clear that the private sector was much more efficient in analysing routine samples. But very special samples are not analysed in the private sector. And the private sector does not develop new technology unless they are paid to do so. But if you [the Rijkswaterstaat, ed.] do not have staff in your lab, you will not be able to contract efficiently, because you don't know the market and the players and they could tell you anything. You always need the basic knowledge in-house, the knowledge necessary to contract'[53].

Eventually, more space for the private sector requires a broader expertise, and expertise of technical nature, than is strictly necessary for good administration and process management; otherwise the private sector may take advantage of the situation:

'The Rijkswaterstaat lost knowledge of the region and the overview of the manner of implementation. In this way much technical expertise on implementation disappeared, leading to a knowledge lack in the Rijkswaterstaat. It is important for a professional and expert outsourcing organisation to have technical expertise to manage the outsourcing process.' (Tjeenk Willink, 2006)

Boonstra en Muijen (2011) indicate that partner organisations such as consultancies and contractors are represented by staff from another age group than their counterparts at the Rijkswaterstaat. Younger employees from the consultancies and contractors have to work together with more senior employees from the Rijkswaterstaat. Similar to the Indonesian case where I described a cultural barrier within the DGWR in Phase III (Section

---

[52] PD70, PD71, PD72, PD73, PD74
[53] PD74

4.3.4.2), we find cultural differences, including differences in working styles and ideas about roles. In the Dutch situation during Phase III, a substantive number of the younger representatives of consultancies and contractors seem not to be interested in the knowledge that can be shared by the senior Rijkswaterstaat professionals. On the other side the seniors are said to be somewhat pessimistic and some of them have lost their enthusiasm for the work. Both attitudes obstruct efficient knowledge exchange.

The survey data drawing on the experience and assessment of a wider group of sector professionals provide a quantitative estimation of the extent of knowledge loss through lack of communication between stakeholders, outsourcing, through retirement, lack of transfer to successor and focus on process management with less attention for substantive knowledge. The data provide an outlook that is consistent with the picture sketched in the interviews. All respondents estimate the loss of knowledge as a result of these five reasons as moderate, with an average score of around 2.7 out of 5, but outsourcing, retirement and a lack of attention for substantive knowledge are perceived to play a growing role in the loss of knowledge from 2000 to 2010 in Phase III as compared to earlier phases (Figure 7.3). A lack of communication between stakeholders is an indirect loss of knowledge, as it is a missed chance for learning and knowledge exchange. Figure 7.3 presents a decrease in the lack of stakeholder communication which is confirmed by the data presented in Figure 7.1. In Figure 7.3 the prognosis for 2015 is optimistic, while in Figure 7.1 this is not the case. The data are furthermore confirmed by data on contact hours of water professionals in the sector (Figure7.8), although I have sampled water professionals only, and the stakeholders would come from a wider background than only the water sector. Albeit the differences between the scores over the years in Figure 7.3 are however not statistically significant, the data may indicate patterns.

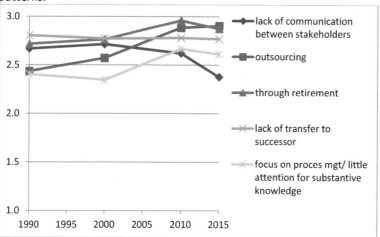

Figure 7.3. Perceived causes of loss of knowledge in the sector, as rated by all respondents (on a scale from 1 = not at all to 3 = to some extent and 5 = extensive)

Outsourcing is also seen as an activity where knowledge potentially is lost, confirming the information gathered from the interviews. In Figure 7.3, as well as in Figure 7.4 and

7.5, the data show an increase from the year 1990 to 2010, (in Figure 7.5 from 2000 to 2010) and a constantly high prognosis for 2015.

In the water sector as a whole, the senior generation of respondents is the most pessimistic in their ideas about knowledge loss (Figure 7.4). However, the younger generation (Figure 7.5) also acknowledges a loss of knowledge through outsourcing, lack of succession planning and a high focus on process management at the cost of substantive knowledge. It is remarkable that the whole group of respondents has become more rapidly pessimistic about knowledge loss in the period 2000 – 2010.

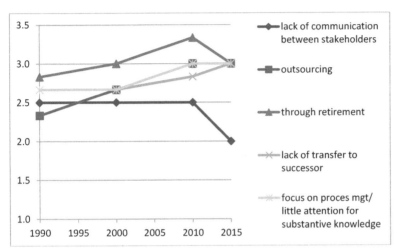

Figure 7.4. Perceived causes of loss of knowledge in the sector, as rated by respondents of 64 years and older

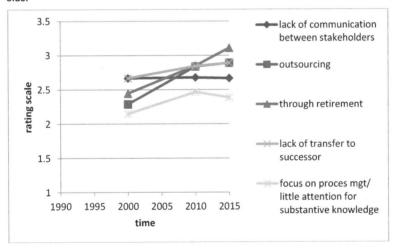

Figure 7.5. Causes of loss of knowledge in the sector, as rated by respondents younger than 39 years old. No data are available for 1990 as most respondents were not yet employed then.

During the years of the reorganisation, various groups discussed the knowledge needs of the Rijkswaterstaat. The documentation of these debates was clustered and analysed in a strategic analysis for use in more structured discussions (Van de Port and Veenswijk, 2006). The analysis made clear that the knowledge needs of the Rijkswaterstaat depend on what the organisation wants to be, which was at that time still subject to debate. It did make the organisation aware, however, of the increasing loss of experience.

The fact that discussion groups are formed, whether these discussions were arranged formally or informally, shows that the organisation has a competence for self-reflection and critical analysis of its own performance, which indicates a competence for continuous learning and innovation (see Table 2.1).

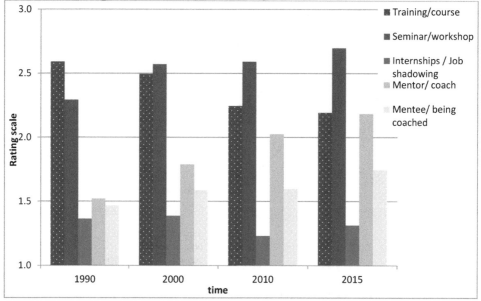

Figure 7.6. Perceived use of KCD mechanisms that need a formal arrangement by the organisation, over the period 1990 – 2010 with a prognosis for 2015 and a rating scale ranging from 1 = not at all, 3 = to some extent, 5 = extensively used.

7.4.5   KCD mechanisms at the organisational level

The survey investigated the role of selected key KCD mechanisms to acquire knowledge and develop capacities. A distinction should be made between mechanisms that need to be arranged and formalised at the organisational level, and mechanisms that staff can use informally on their own initiative. In this section I discuss KCD mechanisms that need to be formally recognised by the organisation in order to be used by staff. I chose a set of KCD mechanisms that are most commonly used in organisations: training and courses, seminars and workshops, internships/job shadowing and the mentor/coach relation.

As can be seen in Figure 7.6 all KCD mechanisms are scored below 3, meaning 'to some extent'. The reason why the scores are relatively low remain unclear, as especially mechanisms such as training, courses and workshops have always been common in most professional organisations. However, respondents likely did not want to overrate any mechanism and this may also be a personality trait. Training is slowly losing ground to other mechanisms such as meetings organised by networking organisations as well as by informal KCD mechanisms that require no facilitation by the organisation, such as accessing online sources of information or better accessibility of data sources in the organisation (Figure 7.7), although it would be expected training will keep a relatively constant role in capacity development of staff. The differences between the scores depicted in Figure 7.6 are not statistically significant at a 90% confidence level.

Internships and job shadowing are clearly not perceived to be used as formal mechanisms to acquire knowledge. Internships are common as a means for (post-) graduate students to gain practical experience in the Rijkswaterstaat, but presumably not for employees to get acquainted with knowledge at other departments.

Networking organisations, e.g. professional associations such IWA or KIVI-NIRIA (see Section 7.1), or the Netherlands Water Partnership (NWP) offer possibilities such as thematic meetings for members, or meetings to get acquainted with other water professionals, which are perceived as becoming more popular over time.

Formal mentoring relationships are important KCD mechanisms for tacit knowledge, as I have explained in Chapter 2. Both being a mentor and a mentee are not rated very high, although their scores are increasing over time. Currently a 'Leaders as trainers' program exists in the Rijkswaterstaat, in which the leaders guide a group of employees (Boonstra and Muijen, 2011). These trajectories make both managers and the group of employees aware of the actual work processes, and how they cooperate. It may not be an exact mentoring relation, but it is a very personal approach to sharing tacit knowledge. In the government specifically, a formal traineeships system has been introduced. The government traineeship is an initiative in which fresh graduates obtain a contract for two years, during which they are guided by a mentor and have easy access to training opportunities. Further ahead in the career this is not common.

For succession planning, mentoring can be an important tool at every level of the organisation. Especially the seniors need to share their wealth of experience with their successors to maintain the institutional memory. This is also pointed out by several interviews[54] and further mentioned in literature and Rijkswaterstaat documentation (Boonstra and Muijen, 2011; Rijkswaterstaat, 2010; Metze, 2008).

Concluding, it proved difficult to indicate to what extent the Rijkswaterstaat has a clear mechanistic or organic organisational structure. From the preceding sections it is clear that the organisation has made a move across Phase I to II from a more mechanistic to a more organic nature and to some extent back to a more mechanistic structure in Phase III. In the first phase the organisation had a relatively rigid task orientation, with a strong hierarchy and obedience to the superior. In the later phases there are elements of mechanistic structures and organic organisational structures. For example, the increased lateral communication after the 1970s and knowledge being available at various levels in the organisation, suggest a tendency to a more organic structure. After the reorganisation in 2002, it seems the measures taken by the management of the organisation to create a more open working atmosphere, as described in Section 7.3.5 were in fact leading to a more mechanistic organisational structure: every activity had to be endorsed by the management, and the structure of control became more strict and hierarchic.

7.5   INDIVIDUAL KNOWLEDGE AND CAPACITY IN THE RIJKSWATERWATERSTAAT

To give an indication of the knowledge and capacities of individual water professionals, I made use of four categories of aggregate competences (see Chapter 2), namely, technical competence, management competence, governance competence and the meta-competence for continuous learning and innovation. In this section I will

---

[54] PD72, PD74

investigate to what extent these three aggregate competences and meta-competence are present at individual level. I will not go into detail on the subcomponents of competence for which I also introduced theory in Chapter 6.

Since the reorganisation in 2002, the Rijkswaterstaat has chosen to summarise its knowledge needs in 5 dominant types indicated with different colours (Van de Port and Veenswijk, 2006; Rijkswaterstaat, 2010):

- Blue knowledge for implementation and operation, mostly substantive technical knowledge;
- Green knowledge about standards and norms;
- Red knowledge about the political environment and society;
- White knowledge on the identity of the organisation, collective memory implying knowledge about the physical system, but also knowledge about the organisation and strategy;
- Black knowledge to be able to determine what should be outsourced and what should be done by the organisation itself; knowledge to control the market.

For a more elaborate description of these 5 types of knowledge, consult Van de Port et al. (2006). The need for knowledge in the Rijkswaterstaat is described in the business plan (Rijkswaterstaat, 2004) as shifting from blue and green, the substantive technical knowledge, before the reorganisation, to red, white and black after. At the same time, the Rijkswaterstaat considered the supply of substantive technical knowledge and expertise a core task of the organisation at the beginning of the reorganisation: this was knowledge for policy support, and to be accountable to the public, to lower government levels, interest groups, to contractors and other actors (Van de Port and Veenswijk, 2006).

The supply of substantive technical knowledge as a core task was stated explicitly in the business plan, but since the start of the reorganisation this core task has been shifted systematically outside the central organisation, notably with the formation of the Delta Institute, now called Deltares. Deltares is a knowledge institute consisting of parts of the former Delft Hydraulics, GeoDelft, the Netherlands Institute for Applied Geoscience (TNO-NITG) and the specialised agencies of the Rijkswaterstaat, namely RIZA and RIKZ. The same has happened for technical substantive knowledge for mobility – a knowledge institute for mobility policy has been established in 2006, a separate institute under the direct responsibility of the Secretary General.

The data for the sector as a whole, and the interview results[55] from the Rijkswaterstaat staff indicate a perceived decreased preference for technical substantive knowledge starting from the year 2000. This indicates that in Phase III the appreciation for technical competence is decreasing, and making way for competence in management and governance. The consequences of the preference for process management at the cost of substantive technical knowledge have been discussed in Section 7.3.5. Literature confirms that the core task of strategic knowledge production slowly retreats, and the Rijkswaterstaat seems to have accepted that for the time being.

An appreciation of red knowledge shows up in the reported increased attention to becoming a public oriented network manager, to have a better understanding of what society needs from the Rijkswaterstaat. The interviews do not provide clues, however, on the increase of this type of knowledge in the organisation.

---

[55] PD70, PD71, PD72, PD73, PD74

White knowledge was also high on the agenda of the business plan, but interviewees indicated[56] that part of this collective memory has disappeared with the early retirement of senior staff. This is knowledge that can party be instilled through special meetings devoted to the identity of the organisation, especially when a new chapter in the life of an organisation begins, but it mostly has to grow by experience, by staying in the organisation long enough and tacitly acquiring this information.

Black knowledge certainly is important considering the 'Market, unless' strategy that the Rijkswaterstaat wanted to follow. I have discussed the pros and cons of outsourcing and its consequences for KCD in the organisation elaborately in Section 7.3.5.

To conclude, it can be argued that the technical aggregate competence has reduced in significance in Phase III to make way for the aggregate management and governance competences. As was seen in Section 7.4 on the organisational structure, it was not anticipated that the management and governance competences had to be recruited from outside the organisation to implement to the new goals and objectives of the organisation, and therefore these individual competences had to be acquired internally after practical experience proved them necessary.

Data on the meta-competence for continuous learning and innovation at the level of individuals was not found.

7.6 KCD MECHANISMS FOR INDIVIDUALS IN THE SECTOR

This section discusses how water sector professionals use informal KCD mechanisms to change or enrich their competences. Informal KCD mechanisms can be defined as the mechanisms that staff can use without permission of their management, that are not part of formal human resource development strategies in the organisation and which require only personal initiative.

Learning by doing scores relatively high for every point in time (Figure 7.7.). On the other hand the importance of a personal network and reading relevant papers and magazines have increased over time. The fact that learning by doing scores high is plausible as much learning happens in an unstructured fashion by trying things out. A similar tendency was seen in the Indonesian case (Section 5.3.4.2).

Figure 7.7 also suggests that the personal network becomes more important for the individuals over time. This can be explained both by developments in the sector, and one's career over the surveyed period. One's network grows over time in the career and fulfils an increasingly important task in acquiring substantive and more general knowledge about one's field, e.g. through information about joint activities. In the water sector in general, the personal professional network has become prevalent as well because society as a whole and the water sector in particular have become more complex.

---

[56] PD70, PD71, PD72, PD73, PD74

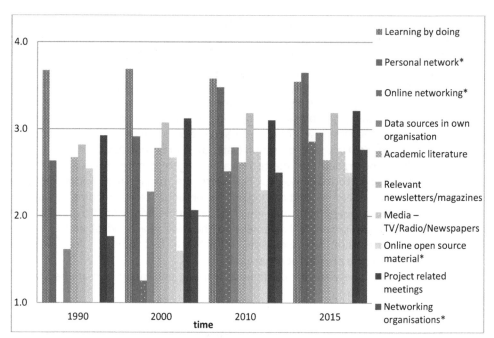

Figure 7.7. Perceived use of KCD mechanisms at the level of the individual over the period 1990 – 2010 with a prognosis for 2015. The asterixes indicate statistical significance at a 90 % confidence level.

Information is offered in many places and it is impossible to keep up to date with all relevant information by oneself and through traditional KCD mechanisms. It is therefore essential to be connected to peers who share their information. In the 1960s and 1970s, networks were important as well, however, the water sector was comparatively simple. The introduction of internet in the late 1990s with its extensive networking possibilities has dramatically increased the exchange of knowledge and information through the personal professional networks.

As can be seen from Figure 7.7, the use of online tools shows a large increase from the start of the new millennium onwards, which is consistent with expectation. The development of internet has provided abundant opportunities for online knowledge sharing, online communities of practice, project websites and online modelling and decisions support systems with easy access for all stakeholders. This is reflected in the increasing scores for online networking and online open source material.

The way of working of water professionals has slowly changed as a consequence of the increasing complexity of water management, and it can be assumed that this is requiring more intensive contact between experts to acquire knowledge. In the survey I investigated the number of contact hours between professionals in the water sector, from 1990 onwards. Contact hours are defined as the amount of time people spend per week in meetings, emailing, on the phone, through social media (after the introduction of internet), informal contacts with colleagues to discuss work and any other means to be in touch with colleagues within and outside the organisation. The results, although not statistically significant at a 90% confidence level, underscore the general trends described in interviews. Figure 7.8 shows a strong growth in the number of contact hours while it is expected that in the future this trend will further continue. The increase may be attributed to the many means available to communicate, the increasing incidence of participation in water management by many different actors since the mid-1990s and the prevalence of team work. Later in Phase II online communication starts to play a significant role. The largest increase occurs in the segment with 16 – 20 contact hours. Interestingly, a number of people indicate that the number of contact hours is 40 or more, which seems quite high.

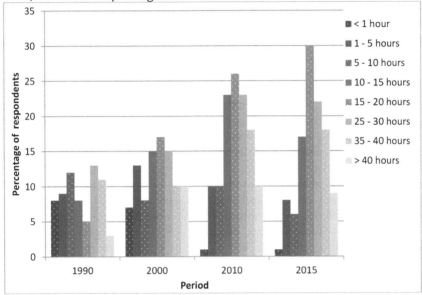

Figure 7.8. Perceived number of contact hours per week between water professionals within and outside the organisation, through e-mail, phone, meetings and other means.

7.7   CONCLUSIONS AND RECOMMENDATIONS

7.7.1   Strategy and method

This case was selected to serve as a reference for the Indonesian case, to compare KCD in the water resources sector in a country where presumably only few KCD mechanisms are available with the situation in an industrialised and richer country, where I assumed more KCD mechanisms to be available. Because of the fact that the Dutch water sector and the transitions in its institutional landscape have been studied in much detail, it was not my intention to do a similar in-depth study as for the Indonesian case. As a consequence I cannot give an elaborate description and indication of the usefulness of all KCD mechanisms, nor can I make detailed statements about the state of the organisation and its knowledge requirements. I succeeded however in describing how the KCD process functions in the institutional environment of the Rijkswaterstaat and to what extent the institutional change over 60 years in the enabling environment led to changes in individual knowledge and capacity.

The mixed method combining semi-structured interviews and a survey with a literature review was used to counteract the weaknesses of either type of method alone (Driscoll et al., 2007). The qualitative tools offered the advantage of improving the understanding of survey responses. The qualitative results partly compensated for the moderate response rate of the online survey and resulting lack of strong statistical significance for some of the results. The quantitative data also provided insight in larger patterns of response. In addition, an adequate amount of literature was available to guide the historical analysis of the institutional framework.

7.7.2 Aggregate competences in the Rijkswaterstaat environment

In this case study I applied the same adapted KCD conceptual framework as in the Indonesian case. The three levels of KCD activity could be identified in a manner that allows comparison with the Indonesian case. The three aggregate competences, (technical, management, and governance) and the meta-competence for continuous learning and innovation (see Table 2.1) also proved workable in this case study as indicators for acquired knowledge and capacity. While the survey results and the qualitative analysis pertaining to institutional and individual competence development yielded useful information, I have attempted to aggregate this information in the groups that we identified in the KCD framework. I have made a ranking order for each phase and each competence group to give an indication of how each aggregate- and the meta-competence have grown or diminished under influence of the institutional environment, the organisational structure and the actions at individual level. I have done this for the Rijkswaterstaat as a whole and for all scales of analysis. On a scale of 1 to 5 (1 = not at all available, 3 = to some extent available and 5 = extensively available) I have scored for each competence in each phase, in comparison with the other phases and competences, and in addition in comparison with the outcomes of the other case study. As a consequence it is a relative ranking, based on my assessment of the phases. The reasoning behind my scores is explained in the text below.

The first phase has been characterised by highly innovative technical development, such as the Delta Works in the South West of the Netherlands. Figure 7.9 shows a high score for aggregate technical competence. Although project management was strong, other types of management, such as natural resource management, were less developed. Therefore, the competence was rated moderate. The cultural identity of the organisation did not change significantly until the reorganisation in 2003. For a long time, the organisation was relatively closed to the outside, only gradually becoming more open in the 1970s, keeping out political interference, with little internal discussion about mistakes. Therefore the aggregate governance competence is rated relatively low in this phase (Figure 7.9.).

Around 1970 champions for nature and the environment in activist groups gained enough political leverage to force the Rijkswaterstaat to incorporate environmental issues in its work. This was the start of a more integrated systems approach to water management. In the development of the Delta Works this led to the establishment of the Delta Agency, a new institute that could focus entirely on innovative design and implementation approaches for the southern Delta Works. The studies of Rand Corporation proved instrumental in providing the external knowledge to start applying a systems approach to water management. At the organisational level this led to increased cross-sectoral communication with other organisations and knowledge institutes, showing a tendency towards a more organic organisational structure. This is in an increased aggregate management competence in Phase II (Figure 7.9). This was taken a step further when in the 1990s, after the floods of 1993 and 1995, the public demanded increased public participation in water management. This forced again the inclusion of a different type of knowledge, namely knowledge of the local stakeholders. Public participation has increased since, but evidence shows that active involvement through co-production and joint-decision making is not very common, partly or probably, because of the inherent complexity and uncertainty regarding responsibility. Until late 20th century, the developments in the Dutch water sector were of a predominant technical nature. This change over Phase II is represented by an increased aggregate governance competence (Figure 7.9.)

The tendency that started already in the 1990s, towards more efficiency and accountability in government sectors, instigated a radical change in the management of the Rijkswaterstaat. From 'Too bulky, too expensive and not transparent', it had to turn into a public oriented and accountable network manager. The sector structure and the Rijkswaterstaat organisation evolved towards objectives, procedures and skills that strengthened governance competence.

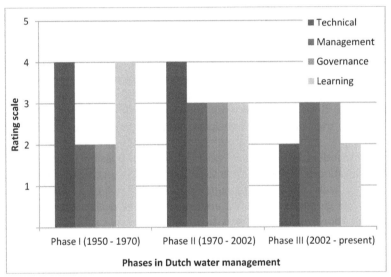

Figure 7.9. Three aggregate competences and the meta-competence for continuous learning and innovation per phase for the Rijkswaterstaat (a scale from 1 to 5, where 1 = not at all available, 3 = available to some extent and 5 = extensively available)

These institutional changes had direct influence on the organisational knowledge and capacity. In the first two phases the organisation moved gradually from a relatively mechanistic position to a more organic position, under pressure of society and politics. The organisation became less hierarchic over time, less control-oriented and its knowledge becomes less concentrated at the top, but spread across the organisation. Lateral communication to solve problems with multiple stakeholders becomes common. In Figure 7.9 this is reflected by a growing aggregate management and governance competence, from Phase I to II. The technical competence stayed at the same level as in Phase I, as the revolutionary technical development of the first phase may have slowed down, but the knowledge was still available in the organisation. In Phase II, the meta-competence for continuous learning and innovation was arguably lower than in Phase I, as, apart from the 'ecological turn' in the beginning of Phase II, the level of learning and innovation is reported not as high as in the first phase. Learning in the second phase mostly consisted of embracing more technical disciplines, which foremost required better management processes and skills.

The reorganisation that started in 2003 brought the organisation into a (temporary) crisis. Due to early retirement of many senior officials and little attention for knowledge transfer, while simultaneously experiencing high staff mobility, much technical substantive knowledge was lost; this is reflected in the decrease in technical competence in Figure 7.9. The HR analysis during the reorganisation showed a decrease in knowledge related to experience. For succession planning, mentorships between senior and junior staff are an important tool to keep the institutional memory in the organisation, but the survey results and interviews do not suggest intensified or dedicated appreciation of this instrument, perhaps rather to the contrary.

The new principle of 'market, unless' and consequent outsourcing of not only the implementation of construction and maintenance, but also part of the preparation, required more knowledge about managing such processes. This knowledge was brought in only later, when it was realised that technical substantive knowledge in the organisation needs to be available too, to be able to control and supervise the outsourced assignments. This led to a gap in technical competence in the Rijkswaterstaat. Governance competence in this phase was supposed to increase as the explicit message from politics was to become more accountable and efficient, and the reorganisation effectuated this. The fundamental question remains whether efficiency has increased by adopting the outsourcing principles. The interviews, survey results and literature shows that this is subject to controversy, and, pending the response of this structure to another demanding situation in an 'acid test', it is not possible to give a clear answer to this. The transition from arrangements determined by technical substantive expertise to ones more driven by process management expertise is a phenomenon that is taking place across the globe, in the public and to some extent also in the private sector. Finding the most appropriate skills mixes increasingly requires careful balancing.

The governance of the Rijkswaterstaat itself in this phase has been subject to criticism. The organisation in this phase turned more control-oriented than before, to achieve more accountability and effectiveness. Every plan, idea or activity in this phase has to be endorsed by the management team; the location of knowledge of actualities is located near the top of the hierarchy, providing less transparency and consequently less trust among staff. The style of governance in the organisation was reported to be less encouraging for learning in this period as staff would refrain from taking initiative and the management team may not have received candid advice from their staff. This is reflected in the lower score for learning competence in Figure 7.9.

The case study shows that the type and number of KCD mechanisms, whether arranged at the level of the organisation or chosen on individual initiative, are of secondary importance compared to the influence the institutional environment exerts on knowledge and capacity. If the institutional settings are favourable for knowledge exchange and capacity development, space will be created at organisational level to establish appropriate KCD mechanisms and for individual staff to take initiative and acquire new knowledge and capacities.

8 Connecting the dots – Discussion and conclusions

8.1 INTRODUCTION

The main focus of my dissertation is the analysis of the dynamic process of knowledge and capacity development (KCD) in the public water sector. To examine the challenges and opportunities for KCD, I chose two entry points. The first entry point was to assess the influence of International Post-graduate Education (IPE) on KCD in the Directorate General of Water Resources (DGWR) in the Ministry of Public Works (MPW) in Indonesia. I hypothesised that IPE is relatively important in a society were few other KCD mechanisms are assumed to be available. Consequently, in the Rijkswaterstaat case, the entry point was the study of KCD in a society where more KCD mechanisms were assumed available. Both cases are examples of large professional water management organisations with a sufficient body of study and information available regarding the important development phases the countries have gone through over 50 years. I used surveys and semi-structured interviews in both cases to analyse how water professionals acquire knowledge and capacities, and I undertook a historical analysis in both Indonesia and the Netherlands to study how the cultural and environmental features and priorities in society at distinct junctures in time influenced the use of certain mechanisms to acquire knowledge and capacity.

I have searched for parallels in the cases in order to infer potential general rules for KCD processes. The adapted KCD conceptual model in Figure 2.1 in Chapter 2 provides the structure for this thesis and is expanded step by step in every chapter towards a revised conceptual model for KCD presented in this chapter. In contrast to other conceptual models (Baser and Morgan, 2008; Robeyns, 2005; Lusthaus et al., 2002) for capacity development I have focused on institutional, organisational and individual knowledge and capacity simultaneously, and the interactions between these three levels, because this provides the best possible assessment of knowledge and capacity in a system. Other models take into account the influence of levels other than the primary level of analysis, but do not assess the existing knowledge and capacity at those levels as well. They focus on the capacity at one level only.

It is not my intention to give evaluative statements about policy choices in both case studies. I make statements about the KCD aspects in the case studies but I cannot judge whether certain policy choices over time were wise or not and whether possibilities were available to make different choices. It would be unwise to compare the Netherlands and Indonesia in such a manner, as both case studies work with very different financial means, very different water management challenges at totally different scales. If I wanted to make evaluative statements I would have to do a more comprehensive study that included investigation of at least all financial means in both organisations, the countries' priorities and decision-making over time.

In this chapter I first of all review the research strategy and methods that I have adopted. Secondly, I compare the findings of the Indonesian and Dutch case study, while at the same time answering the research questions. Then I explain the scientific and development contributions that go further than the research questions. The chapter will end with a number of recommendations for practice.

8.2 REVIEW OF THE METHODS

The strategy to undertake a historical longitudinal analysis in both cases yielded rich insights. It introduced a very useful differentiation of the key variables and as such it avoided the need for many cases as it reveals, uncovers, and discusses parameters that would otherwise have to be found in other cases. It would however have been a valid option to research more cases, but without a longitudinal analysis in each case. The choice for the two cases also proved to be productive with respect to the many parallels that were found through comparison of the cases.

The Indonesian case took much more time because historical analyses were not available yet, and obtaining reliable information was more difficult than in the Netherlands, because the researcher had to adapt to Indonesian culture.

I adopted a mixed method design for both cases to offset the weaknesses of either approach alone (Driscoll et al., 2007), consisting of semi-structured interviews and a quantitative survey. The mixed method design has provided advantages because the qualitative data provided a deep understanding of survey responses, and the quantitative data provided insight in larger patterns of responses.

The methodological differentiation of respondents in the Indonesian case as a function of their LPE or IPE experience, and for both cases the differentiation of the administrative and political contexts in the country and sector per phase proved useful in generating detailed insights in the development of the competences in the Indonesian and Dutch water sector, over a long span of time, within the evolving economic, administrative and political contexts.

The combined utilisation of guided surveys, semi-structured interviews and the analysis of reports and policy papers proved essential in ensuring accurate and meaningful interpretation; survey results alone were often insufficient and hard to interpret. Importantly, surveys reflect the perceptions of individuals, and as many respondents in the Indonesian case have had only one and a very personal experience with LPE or IPE and can therefore not compare, these perceptions are not necessary mutually compatible; the presence of the interviewer/analyst in the Indonesian case offered the possibility to set more robust assessment benchmarks. This was however very time consuming, and the presence of the researcher may lead to unwanted influence on the answers of the respondents, and an interviewer bias in the results. To overcome the problem of reliability, I have triangulated the data with qualitative data from the semi-structured interviews and literature in both cases.

Furthermore, Atlas TI proved useful in structuring the qualitative data analysis process. Coding could be done in a systematic manner, providing rigour to the analysis process, and increasing the transparency and reliability of my research process. It is however time and labour intensive.

8.3 REVIEW OF THE CASE STUDIES

In this section I will discuss the substance of the two case studies and compare the cases to each other. I will answer the research questions in each of the following sections:

1. How does the institutional environment influence the development and use of knowledge in the public water sector?

2. How does the organisational structure influence the development and use of knowledge in the public water sector?
3. What KCD mechanisms are available at the broader institutional, organisational and individual levels, and to what extent are they used?

Table 8.1 and Figure 8.1 and 8.2 synthesise the findings of the study of the Dutch and Indonesian case to facilitate the comparison. In Table 8.1 I present the most important characteristics of the phases per level of analysis. Figure 8.1 and 8.2 show a ranking order for each phase and each aggregate- or meta-competence to give an indication of how the competence has grown or diminished under influence of the institutional environment, organisational structure and individual knowledge and capacity in each phase. I use Table 2.1, indicating the three aggregate competences and meta-competence for continuous learning and innovation at each level of analysis as a guideline to indicate the level of competence.

Using a Likert scale of 1 to 5, where 1 = not at all available, 3 = available to some extent and 5 = extensively available, I have ranked each competence in each phase. I have ranked by comparing with the other phases and the other competences and given an average score for the institutional, organisational and individual levels, and in comparison with the other case study. As a consequence it is a relative ranking, based on my understanding and assessment of the competence present over the phases in each case study.

8.3.1 The influence of the institutional environment on KCD in the public water sector in Indonesia and the Netherlands

Chapter 4 presented an historical longitudinal analysis of the Indonesian water sector the institutional environment for policy change and its relation with KCD in the DGWR, from the 1970s until the present. In Chapter 7 a similar analysis, albeit more brief, was undertaken for the Dutch water sector, in relation to KCD in the Rijkswaterstaat, from roughly the 1950s until present. In both case studies I consider the water sector to be confined to water resources management. To be able to demonstrate how the changes in the institutional environment influence knowledge and capacity at the broad institutional, organisational and individual levels, I divide the history into distinguishable phases, based on shifts in paradigms.

In the Indonesian water sector, Phase I was a period remembered with pride, because of a steady economic growth and the focus on the construction of infrastructure (Table 8.1). It was imperative to build up transportation, communication and water infrastructure (flood protection works, hydropower, irrigation and water supply) all for economic development and food security.

The predominant institutional paradigm and the advocacy coalition in the DGWR was a construction-oriented, technocratic coalition, built on the belief that infrastructure development was important for the nation's development, and that it was best to keep control at the central level, as only DGWR had the capacity to plan and implement this.

Table 8.1. Comparison between the Dutch and Indonesian case study

Level of analysis	The Netherlands			Indonesia		
	Phase I (1950 – 1970)	Phase II (1970 – 2002)	Phase III (2002 – present)	Phase I (1970 – 1987)	Phase II (1987 – 1998)	Phase III (1998 – present)
Institutional level						
Political shifts and dynamics	Infrastructure dev't to support growing economy	Environmental values included by left-wing govt	Privatisation - increased attention to transparency and efficiency of government	Infrastructure dev't to support growing economy and rice self-sufficiency -	Increasingly authoritarian state -	Reformation leading to decentralisation - increased transparency and accountability
Main legislation and policies in place	1st National water policy (1968): increasing water supply with large scale infrastructure development	2nd National water policy (1985): economic approach to water supply + water quality, 3rd national water policy (1989): water quality and quantity, 4th national water policy (1998) based on IWRM, European Water Framework Directive (2000)	Water act (2009), preparation of environment act (2013), the National Water Plan (2009)	Water law 11 (1974), government regulations on water resources management (1982), irrigation (1982)	Government regulations on swamps (1991) and rivers (1991)	Law 7 on IWRM (2004), government regulations on water supply & sanitation (2005), irrigation (2006), groundwater management (2008) water resource management (2009), dams (2010)
Organisational culture	Technocratic- hierarchic and closed - strong esprit de corps – little space for other professions than engineering	Space for other professions than engineering - increasing role of civil society and stakeholder participation -	Increasing use of internet for knowledge exchange and networking – after reorganisation more control oriented	Technocratic - strong esprit de corps - little space for other professions than civil engineering - high power distance on the work floor	Little space for other professions than civil engineering -loyalty increasingly important – ongoing high power distance on the work floor	Increasing openness and transparency - through internet better accessible information leading to more transparency – power distance slowly diminishing
Organisational level						
Mechanistic vs. organic	Mechanistic: hierarchic - precise division of tasks	More organic: Increased lateral communication with colleagues and other organisations - substantive knowledge can be located anywhere in the organisation	More mechanistic: hierarchic- and knowledge located at the top of the organisation	Mechanistic: hierarchic divisions - precise division of tasks - communication of instructions and decisions from the top	Mechanistic: Little coordination with other departments or organisations – hierarchic - loyalty and obedience	Slowly becoming more organic – more information available more lateral links

KCD mechanisms established by organisation	Project related meetings – training - workshops	Project related meetings – training - workshops	Seminars/ workshops	downwards Training and education - seminars and workshops - formal meetings	Seminars - formal meetings	Training and education – seminars and workshops – formal meetings -
Individual level						
Important competences	Focus on internal technical knowledge – competence for learning	Inclusion of other technical professions - introduction of IWRM - stakeholder involvement - attention for external interdisciplinary knowledge	Process – public relations – slow reacquisition of technical substantive knowledge	Focus on internal technical knowledge	Ongoing focus on internal knowledge - erosion of technical expertise	More attention for management competence and governance competence: management of outsourcing, stakeholder involvement, policy formulation skills
Informal KCD mechanisms	Learning by doing – informal meetings	Learning by doing – informal meetings – personal/professional network	Increasing use personal/professional networks through the whole sector under influence of internet – learning by doing	Informal meetings in the organisation – mentoring relations - learning by doing	Informal meetings in the organisation – learning by doing	Learning by doing – increasing use personal/professional networks under influence of internet – informal guidance by seniors

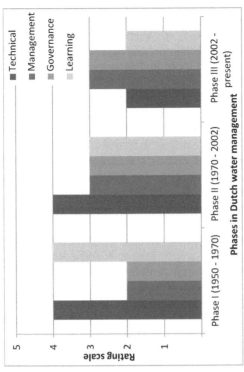

Figure 8.2. Aggregate competences per phase for the Rijkswaterstaat (a scale from 1 to 5, where 1 = not at all available, 3 = available to some extent and 5 = extensively available)

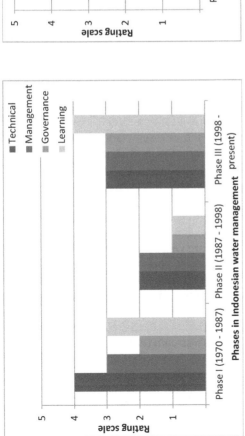

Figure 8.1. Aggregate competences per phase at for the DGWR in Indonesia (a scale from 1 to 5, where 1 = not at all available, 3 = available to some extent and 5 = extensively available)

149

There was little participation from other stakeholders, except from the international donor community. In this period the institutional setting (Table 8.1) was appropriate for the goals of infrastructure development for economic development and food security, facilitating the appropriate organisational and individual capacity as well. For example, the technical expertise and skills mix was available (Figure 8.1), and was relatively straightforward, knowing that the main goal was to design and build infrastructure. Furthermore, institutional authority for the conceptualisation, financing, construction, operation and maintenance, normally a complicated network of task assignments was kept very simple: DGWR was responsible for everything. What was lacking in terms of institutional competence was the inclusion of various stakeholders in any part of the process, and the presence of an entity responsible for operation and maintenance (O&M) and cost recovery, which is reflected in a relatively low score for the aggregate governance competence (Figure 8.1). The centralisation was a rational policy choice, necessary in an environment with only little infrastructure and with little technical capacity at regional level. However, it should be noted that this centralised model was common across the world in this period and only in the 1970s and 1980s would a few countries start introducing more decentralised arrangements, notably for O&M.

In the Dutch water sector, the period until the 1970s was dominated by feelings of pride, because of the major achievements in construction (Table 8.1). In the Netherlands the focus was on reconstruction after the Second World War, and on flood protection. The masterpiece of the Rijkswaterstaat after the storm surge of 1953 –was the construction of the Delta Works. The organisational culture (Table 8.1) - the strong esprit de corps, the closed front towards the outside and the belief in its own technical superiority – show many similarities with the Indonesian case. Learning concentrated on technical substantive knowledge. Lintsen et al. (2004) mention that conflicts within the strict hierarchic organisation of the Rijkswaterstaat would never come out, which is similar to the early culture in the DGWR. During the reorganisation of the Rijkswaterstaat in 2002 a control oriented culture was developed, that shows resemblance to the Indonesian case in the sense that the power distance between staff and superior became larger and communication was relatively top-down. Employees felt there was little space for disagreement, and they started to take less initiative, afraid to stand out. A similar atmosphere is described in the Indonesian case.

The differences between the two cases at the institutional level lie in the way the advocacy coalitions managed to argue their agendas, and in the political leverage. First, in the Netherlands, civil society organisations are well funded and often subsidised by the government, but may still criticize the government's performance, if they feel inclined to do so. These organisations played a significant role in creating awareness on environmental issues in the 1970s, and managed to gain the political support of the centre-left wing government of Prime Minister Den Uyl. Externally visible events helped to create awareness, such as the Rhine pollution that left fish dying at the river surface. The political stream (Den Uyl) and the problem stream (dead fish) (Kingdon, 1995) came together and provided enough leverage to change the existing policy. In the Rijkswaterstaat the 'ecological turn' (Disco, 2002) instigated under growing professional political and societal pressure, led to the creation of the Delta Agency, the agency responsible for the design and implementation of the Southern Delta Works. The Delta Works were a response to the storm surge in the South-West of the Netherlands in 1953, to protect the hinterland against floods. Enough institutional capacity was available to ensure that this organisation was established, and also to create a culture in which knowledge creation and exchange flourished and where it could be acknowledged that certain competences had to be brought in from outside sources. This is reflected in an increased aggregate competence for both management and governance (Figure 8.2). For example, with the help of external knowledge brought in by Rand Corporation, i.e. the American think tank that introduced systems thinking in Dutch water management, the foundations were laid for Integrated Water Resources Management (IWRM).

In the Indonesian case, knowledge brought in by the donor community in the 1970s – 1980s was perceived as threatening to the status quo in the DGWR, just like the Rijkswaterstaat felt threatened by the environmental lobby, but in the Indonesian case the donor community did not manage to change the existing technocratic water policies of Phase I, because there was no political leverage for IWRM and more decentralised decision-making. The institutional competence for inclusion of more stakeholders in decision making was not yet available. Both IWRM and the Irrigation Operation and Maintenance Policy (IOMP) require the surrender of power to more stakeholders and a shift from a building to a management paradigm. As the regime was becoming more authoritarian in Phase II, and construction remained lucrative, inclusion of other stakeholders was not high on the agenda. Knowledge on IWRM and on Operation and Maintenance (O&M) was contested. Knowledge can play an effective role in policy processes, that is, if it fits with the paradigm of that moment. The political situation thus did not provide a window of opportunity for significant policy change; the change from Phase I to II in Indonesia was only superficial, and not a deeply embedded paradigm shift.

In Indonesia the shift from Phase II to III provided the window of opportunity to shift to a new paradigm. The fall of the political regime in 1998, opened the way to a far-reaching decentralisation policy, driven by strong public pressure. This led to an increase in the aggregate competence (Figure 8.1) in the DGWR shown in the establishment of a National Water Resources Council in which different stakeholders have seats, decentralisation of responsibilities and more openness to assess the department's performance. However, there is still a clear hierarchy of knowledge, as is often the case in situations where technical expertise is the monopoly of a single profession - in this case engineering - and this is reinforced in the relatively closed community of the DGWR. Technical knowledge as a basis is very much appreciated, supplemented by knowledge to run the administrative processes. Similarities with the Dutch case in this phase are found in the increased attention for accountability and transparency, and the extensive outsourcing under influence of the development of the policy ideas of the Washington Consensus in the early 1990s. In both cases the organisation is struggling to manage the outsourcing process because of a lack of technical competence (Figure 8.1 and 8.2) in the organisation itself.

To conclude, both cases show that the aggregate competence at institutional level is a necessary condition for aggregate competence at organisational level and in turn provides the space for knowledgeable individuals. In turn, knowledgeable individuals have to organise themselves into coalitions, help create and make use of the momentum (the merging of the problem stream, political stream and policy stream) to create sufficient conditions for capacity development and pave the road for new knowledge and capacities.

8.3.2   The influence of the organisational level on KCD

Organisational competence is not assessed in this thesis by quantitatively investigating the number of staff with certain competences to come to an indication of the composition of the organisation, and statements about the type of T-shaped competence profiles in the organisation will not be made. Instead I provide a qualitative assessment of organisational competence. To do so, I adopted Burns and Stalker's characteristics of mechanistic and organic formal organisational structure (1961) and compared the DGWR and the Rijkswaterstaat with these characteristics. I chose to work with this theory because the application of the characteristics of mechanistic and organic organisational structure to the DGWR provided insight in the flexibility of the system in finding solutions to changing conditions and complex challenges. These challenges may consist of the inclusion of environmental objectives and approaches in water management or the shift to modern management and organisation process, which cannot be handled without engaging with stakeholders outside the organisation as well.

No recipe exists for the extent to which an organisation needs to work with an organic or mechanistic organisational structure. Government organisations tend to have a more mechanistic organisational structure because the work consists partly of routine tasks for which such a structure is appropriate. Organisations that function as think tanks or innovators need a much more organic organisational structure, because for creativity people generally need an environment free from rules and regulations. The more pressing the need for external knowledge, the more organic the formal organisational structure needs to be, to be able to efficiently access that knowledge. This does not mean that a quick but smart decision by the top management is not possible or sometimes necessary anymore. An organic formal organisational structure places a heavier demand on individual competences, requiring intensive communication and extensive discussion with peers. The organisation must facilitate this intensive communication in and across the borders of the organisation, enlarging its own organisational competences such as the development of personnel, and leadership skills. In doing so, the organisation provides the people working in these organisations with the authority they need to respond to the conditions they encounter (Rainey, 2003).

Coordination and communication with other organisations, were not common in both cases in the first phase, as reflected in the relatively low score for the aggregate governance competence in Figure 8.1 and 8.2. In the shift from Phase I to Phase II, the Rijkswaterstaat had to become more organic as a result of political pressure to solve growing environmental problems, and accept a wider array of professions and knowledge inside its 'walls'. Especially in the Delta agency, space was created for elaborate discussions and interdisciplinary cooperation, in a much less hierarchic environment than the Rijkswaterstaat employees were used to (Table 8.1).

Similarly in the Indonesian case, in the shift from Phase I to Phase II the organisational structure had to become more organic if the organisation wanted to allow international knowledge about IWRM to be accepted inside the organisation. Differing from the Dutch case, this could not happen because the political climate outside and in the organisation tended to reward loyalty to interest groups more highly than professional competence and integrity, again showing that the formal organisational structure of an organisation is a reflection of the institutional context. Consequently, after the transition to Phase III, the administrative system continued to reflect the institutions that partly remained from the Suharto regime, and partly are culturally defined. In Phase III, the DGWR was still marked by a high degree of formalisation and centralisation of responsibilities, expressed in a strong hierarchy, and strict division of labour and routines. Even though water management challenges increasingly demanded interdisciplinary knowledge and capacities for resolution, meaning the acceptance of knowledge other than only engineering, and from external sources, the formal organisational structure is still relatively mechanistic. Part of the reason could arguably be the weak voice of civil society in demanding quality service delivery. Another reason may be that coordination and collaboration with other parties that typically tend to be comparatively mechanistic do not sit well with a control-oriented organisation such as a ministry responsible for water management.

Therefore, I conclude that the management competence in the Indonesian case is relatively weak in Phase II (Figure 8.1), based on how the formal organisational structure has changed little in that period time, while the existing water management challenges were pressing for the inclusion of more diverse sources of knowledge. In Phase III with an emphasis on decentralisation, more transparency and accountability, the organisational structure slowly starts showing characteristics of a more organic structure. As the power distance in the organisation is slowly diminishing, the organisation will increasingly provide space for the location of knowledge at various levels in the organisation and not only at the top.

In the Dutch case, organisational competence was higher in the second phase (Figure 8.2) as the Rijkswaterstaat adjusted to the pressures emanating from the institutional level and became more organic, providing space for lateral communication, frequent contact with other disciplines and organisations and openness to external knowledge, facilitating the growth of individual competences as well. Through every transition the organisational structure has had to become more organic once more to facilitate an increasing array of sources of knowledge and to facilitate the exchange of knowledge in the organisation, to encourage employees to take initiative, solve problems and feel ownership for the work of the organisation and to grow in competence. In the Dutch case this succeeded best in the beginning of the second Phase (Figure 8.2).

8.3.3 KCD at the individual level

To visualize the combination of the essential aggregate competences that were part of the initial adapted KCD conceptual model, I applied the theory of Cheetham and Chivers (2005) combined with the theory on the T-shaped competence profile, developed by Oskam (2009). This proved helpful in distinguishing the different subcomponents of competence, namely cognitive explicit competence, cognitive tacit competence, functional competence or skills, and personal/ethical competence. Often only factual knowledge and skills are considered, but personal and ethical competence are just as important for successful performance. The subcomponents of competence are very similar to the components of knowledge as defined in the Theory Chapter. The division in subcomponents furthermore provides clarity about the type of KCD mechanisms that are most suitable to develop the competence.

In the Indonesian case, the survey of competence formation by individuals shows that the DGWR and the sector professionals have a firmly 'technical' default orientation, especially when individuals choose their field of education. However, they also express a strong need for the other aggregate competences and meta-competence: primarily the meta-competence to learn and innovate, as well as the aggregate competence for management. Although the default orientation may be technical, concerns are now being expressed that especially line civil servants have a strong preference for administrative skills, and that the balance with substantive technical competence is lost. As most staff prefer to be(come) a line civil servant, the balance may tip to a situation with too little substantive knowledge.

Interestingly, a similar development was observed in the Dutch case where the reorganisation of the Rijkswaterstaat of 2003 to 2007 led to a situation where process management was emphasised and where the substantive technical knowledge, on implementation and operation and standards and norms, was perceived as underappreciated (Table 8.1).

In both cases this created a situation where it became more difficult for the departments to prepare, initiate, supervise and control outsourced work.

In the Indonesian case, the aggregate competences for management as well as governance were reportedly insufficiently addressed, in the main channels that are utilised to acquire professional knowledge, whether through LPE or IPE, even though respondents indicate that these competences are important. With little competence in these fields, it will be difficult for professionals to communicate and collaborate effectively in interdisciplinary settings and in the water policy and political discussions that characterize the Indonesian water sector since the onset of Phase III. This may limit civil servants in effectively influencing policy processes or becoming true policy entrepreneurs. The substantive knowledge is necessary to feed into the policy process, and knowledge on governance is necessary to communicate with various stakeholders and to know how to influence that policy process. In the Indonesian case, the competence to effectively formulate policy or influence policy making by forming coalitions is not acquired from local or international post-graduate education. This is clearly a competence that is acquired through learning by doing. Since the alumni from international post-graduate education return with new substantive knowledge that would be useful for the organisation, it may be advisable to pay explicit attention to the policy formulation skill. At the same time the professionals studying abroad are becoming younger and younger, and upon return to their home organisation they will not yet be in a position where they can exert much influence on policy processes.

In both cases it has become very clear that competence at individual level is directly related and nested in the organisation in which a professional works and its institutional environment. Competences of the organisation, e.g. Human Resource Management (HRM) systems, and the institutional environment, e.g. a culture that fosters learning, a sense of urgency to work on water management challenges; have direct impact on the individual competence, be it technical competence to design waterways, governance competence to know how to design in a participatory manner, or the meta-competence for continuous learning and innovation and to critically reflect on one's performance.

8.3.4 KCD mechanisms and the role of tacit knowledge

The extent to which particular KCD mechanisms are employed either those formally arranged by an organisation or those used by individual staff informally is of interest in this study. However, in the discussions about KCD mechanisms during interviews, in both cases more emphasis was placed on tacit knowledge gained through the mechanisms rather than on the mechanisms themselves.

First of all, in both cases personal professional networks are reported to be important, albeit for different reasons. In the Dutch case, personal networks have become more important because society as a whole and the water sector have become more complex. It is important to be connected to peers, who share information about initiatives, proposals and events, because it is too time consuming to attempt to gain an overview by yourself. The personal professional network is moreover important to advertise and highlight one's own activities. The use of the personal professional network has enormously increased with the help of online tools for networking and by the social media. In the Indonesian case the personal professional networks are just as important, but mostly because information provision in the organisation is incomplete.

Second, apprenticeships were perceived to be important in the Indonesian case. In the Indonesian case many interviewees and survey respondents indicated mentorships in Phase I as an important KCD mechanism, by which the junior engineers gained understanding of practical applications of their theoretical knowledge quickly. It became less common in Phase II and III. In the Dutch case it remains unclear whether formal or informal apprenticeships are common. Among young employees who follow a traineeship in the organisation it is common for the duration of the traineeship. In both cases it would be advisable to institutionalize the mentor – mentee relationship, as it is a well-known mechanism (Nonaka and Takeuchi, 1995; Vernon and Werner, 2010; Lave and Wenger, 1991) for transferring and exchanging tacit knowledge and to passing on the institutional memory from seniors to younger staff. Both cases furthermore report of cultural barriers in knowledge exchange between seniors and juniors that could be diminished if mutual understanding was created in a mentoring relation.

Third, Chapter 6 on competences acquired through post-graduate education, indicated that the tacit knowledge acquired during IPE, such as the exposure to a different working culture and different learning format was fundamentally formative. Respondents indicate this to be the most important aspect of their international education, even though this does not get explicit attention in the curricula themselves. This clearly shows the importance attached to tacit knowledge by water professionals.

At the same time both organisations lose tacit knowledge. First of all, succession planning requires serious attention as one of the means to keep tacit knowledge inside the organisation. In the Dutch case, the reorganisation that started in 2003 has resulted in a loss of institutional memory and experience. Knowledge has been lost because of the early retirement of many senior officials, and despite the high staff mobility, little attention was paid to knowledge transfer. A strategic personnel analysis during the reorganisation showed a decrease in knowledge related to experience, which is effectively tacit knowledge. Over the last few years this situation has received more attention. In the Indonesian case the issues are similar, albeit more urgent, as the number of senior professionals retiring in this coming decade is dramatically high.

Second, in both cases respondents and interviewees indicated that a substantial amount of work is outsourced, and both organisations struggle with the question of whether they should outsource not only implementation and maintenance but also the preparation of projects. In both cases it became clear that to be able to supervise the contractors and to obtain a quality product, the organisation itself needs more capacity, to control the outsourcing process from beginning till end. In both cases this capacity is not available anymore as it was tacit knowledge in the mind of engineers that have retired. Since in addition to substantive technical (tacit) knowledge about the assignment that can only be obtained by executing part of the work in force account, administration and process management skills are imperative to successfully manage the outsourced assignment.

To conclude, tacit knowledge needs to be addressed explicitly in organisations, in formal KCD mechanisms such as education and training, by arranging succession planning, by providing the opportunity to enter a mentor-coach relation, but also by creating more informal opportunities and an atmosphere for knowledge exchange.

8.4   REVIEW OF THE THEORETICAL FRAMEWORK

8.4.1   The adapted KCD conceptual model

In Chapter 2 I explained the building blocks of the adapted KCD conceptual model as a basis that offered a comprehensive overview of knowledge and capacity at the level of the individual, organisation and institutions and of the appropriate KCD mechanisms that are typically applied to acquire knowledge and develop capacity. The adapted KCD model takes into account the dynamic institutional environment, but also identifies components that should be assessed to give an indication of the capacity at this level. It identifies how capacity at this level influences the functioning of the KCD at the 'lower' nested levels, and vice versa. The three levels contain their own types and extents of knowledge and capacity and the three levels are in dynamic interaction, they mutually influence each other.

The adapted conceptual framework of KCD has furthermore been used as an ordering framework for this thesis, analysing one level per chapter in-depth in the Indonesian case and for the smaller Dutch case a complete assessment across the levels in one chapter.

The adapted KCD conceptual model in Chapter 2 does not provide direct insight into the relationships between the levels, nor between existing knowledge and capacity and KCD mechanisms, nor does it indicate how the aggregate competences and the meta-competence serve as indicators of the success of a KCD intervention. Instead complementary theories were required at each level of the conceptual model to deepen understanding of these connections.

The distinction of three levels of KCD in the adapted conceptual framework has consequences for the choice of complementary theories. For example at the organisational level, the distinction forces me to search for a theory that investigates the formal organisation only, as I have defined the informal arrangements as institutions. A distinction between the formal and informal workings of an organisation is somewhat artificial as in practice people make use of formal and informal arrangements interchangeably and when deemed appropriate for their situation. If I do not separate the formal and informal rules and arrangements, I can only observe and describe phenomena, typified by the bricolage concept of Francis Cleaver (2002). In seeking to measure and compare, I am obliged to move beyond this stance. Separating the levels has had a positive influence on the results as it has enabled the clear distinction of the formal organisational structure as a prominent factor that itself determines KCD.

In addition to the adapted model described in Chapter 2, I have conceived a complementary depiction that illustrates how the theory drawn upon in this dissertation deepens understanding of knowledge and capacity development (Figure 8.3). The new depiction reflects the dynamic nature of the interactions between the different levels, and notably how the organisational level co-defines KCD mechanisms at individual level, and how the KCD mechanisms of both the organisational and individual levels are affected by the institutional environment level, in which they are embedded, which was less clear in the adapted KCD conceptual model. The new depiction can be viewed as complementary to the adapted model, highlighting the dynamic, nested effects of the individual, organisational and broader institutional layers upon each other.

8.4.2 Four components of professional competence

At the level of the individual, I have introduced theory on competences, as described in Chapter 6. It was necessary to not only look at the aggregate competences and meta-competence for continuous learning and innovation, but to further break down the concept of competence into different components. The different components, i.e. cognitive – explicit competence, cognitive-tacit competence, functional competence and personal/ethical competence provide information on which KCD mechanisms to use to improve the component, e.g. for the cognitive tacit components, a mentor-coach relation will be a better option than classroom education, whereas for functional competences, on-the-job training is appropriate. Depending on the organisational structure and the institutional environment, particular KCD mechanisms can be used or not (Figure 8.3).

8.4.3 Mechanistic and organic formal organisational structure

Burns and Stalker (1961) described the characteristics of mechanistic and organic systems. A mechanistic structure is appropriate for stable conditions whereas an organic structure is viewed as more appropriate for changing conditions. Whether the organisational structure is more mechanistic or organic is amongst others determined by investigating the distribution of tasks, task scope and conformance and the structure of control and authority in the organisation. If the formal organisational structure is relatively organic, it provides room for different KCD mechanisms than when it is more mechanistic. A mechanistic organisational structure often can provide for training and education, more traditional mechanisms for collective KCD at the level of the individual, but more personal KCD through coaching, apprenticeships, as informal lateral communication with professionals outside one's own department or organisation is less common. At the level of the organisation, KCD mechanisms such as change management or organisational learning will not receive much attention in a more mechanistic structure. Organisations with a more organic structure will be more open to adapt the roles and procedures in the organisation to manage the challenges that arise. The theory is extensively described in Chapter 5.

8.4.4 The Multiple Streams Framework and Advocacy Coalition Framework

As is further indicated by the nestedness of the organisation in its institutional environment (Figure 8.3) the way the organisational structure is designed depends on the institutional environment in which it operates. This was investigated in this thesis using the Multiple Streams Framework of Kingdon (1995) and the Advocacy Coalition Framework, worked out by Sabatier (1993; 2007) (Figure 8.3). This theory helped to generate better insight in how personal knowledge is used in an organisation in order to come to policy change and for change in the institutional framework. Actors (e.g. donors, change agents in the organisation or others) that intend to influence the direction of an organisation generally need to form coalitions with others to bring about change. As explained in Chapter 4, this change will most often occur if momentum is created because an issue is brought to the political agenda by means of advocacy or because external events take place that create forces and incentives to shape the new political and policy preferences, and the organisation is ready to work along these lines. In other words, the problem stream, policy stream and political stream need to merge to create a window of opportunity for policy change. In Figure 8.3, the problem stream is drawn partly outside the institutional environment and partly inside, indicating that it can be fully external, or a problem emerging from within the institutional environment of the individuals and organisation. Policy change in turn leads to changes in the KCD policies and the appreciation of a new combination of aggregate competences at each level, as depicted in Figure 8.3. Civil society can be represented in an advocacy coalition, but is also an actor on its own. For example in the Netherlands in the 1970s, society became increasingly aware of environmental pollution and this sentiment was picked up by ruling political parties, leading to large scale changes in the way the country managed its water resources. In Indonesia, in 1998, the Suharto regime fell because the population was dissatisfied with the increasing corruption and the self-enrichment of elites, leading to a new political regime advocating more transparency and accountability to the public.

8.4.5 A new depiction for KCD

The adapted KCD conceptual model (Figure 2.1.) was a comprehensive ordering framework and functioned well as such. Figure 8.3 shows a broader framework that positions the different theories according to the help they offer in the operational analysis and describing of the key processes in KCD. The adapted model and the new depiction can be used together to assess how knowledge and capacity are developed at each level, how the levels interact and how knowledge and capacity flow between the institutional, organisational and individual level. The new depiction therefore represents an expansion of the original concept, or a deepening of the framework to provide the analytical tools for studying the dynamic interaction between the three levels, and in particular the key role the institutional level plays. For readability, the KCD instruments at each level are not listed explicitly.

The individual level is displayed as being nested in the organisation, which is in turn nested in the institutional environment, indicating that the levels mutually influence knowledge and capacity at the other levels. The various advocacy coalitions are displayed partly inside the organisational sphere and partly in the institutional environment, as they can consist of individuals in and outside the organisation, but always individuals that have an interest in changing the policy of that organisation, and therefore firmly inside the institutional environment. In the presentation of the three streams of Kingdon, I have chosen to depict the problem stream partly outside the institutional environment, as a problem such as flooding for example may partly happen because of excessive rainfall, an external cause, but it may be aggravated as a result of ineffective governance in the river basin, which is part of the institutional environment of the organisation. The arrows in the figure show the directions in which influence is exerted.

The momentum created when the three streams merge temporarily, can lead to a window of opportunity for new advocacy coalitions to consolidate their ideas in new policy.

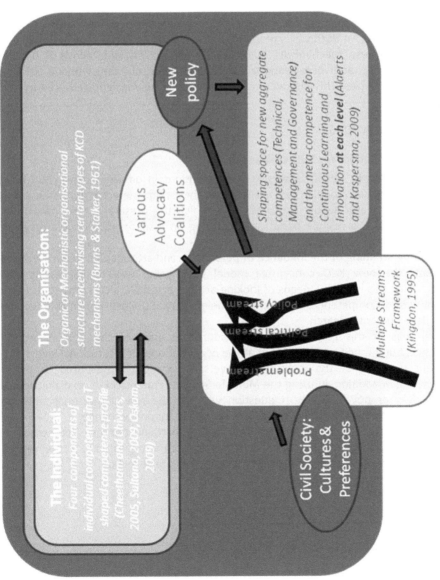

The Organisation:
Organic or Mechanistic organisational structure incentivising certain types of KCD mechanisms (Burns & Stalker, 1961)

The Individual:
Four components of individual competence in a T shaped competence profile (Cheetham and Chivers, 2005, Sultana, 2009, Oskam, 2009)

Various Advocacy Coalitions

New policy

*Shaping space for new aggregate competences (Technical, Management and Governance) and the meta-competence for Continuous Learning and Innovation **at each level** (Alaerts and Kaspersma, 2009)*

Multiple Streams Framework (Kingdon, 1995)

Problem stream

Political stream

Policy stream

Civil Society: Cultures & Preferences

Figure 8.3. A new conceptual model for capacity development at the individual, organisational and institutional level

159

8.5   SIGNIFICANCE OF THIS STUDY

This study has increased the scientific rigour in the field of knowledge and capacity development and has demonstrated that the elusive concept of KCD can be analysed using a structured research approach. It has further demonstrated a workable method for measuring the success of KCD mechanisms.

The historical analysis was useful in clarifying the role of new knowledge through the years and the influence of external triggers and political will. Interesting similarities and differences between the two cases were identified, from which we may conclude that the mechanisms that determine KCD in both cases are fundamentally similar.

The first and the second research question could be answered satisfactorily in this research as described in earlier sections.

For the third research question, the available KCD mechanisms have been identified and researched. However, in both cases more emphasis was placed on the tacit knowledge gained through these mechanisms, and which mechanisms are 'allowed' by the institutional climate and organisational structure, than on the working of the mechanisms themselves. This discovery can be seen as a new insight delivered by this research.

8.6   RECOMMENDATIONS

8.6.1   Recommendations for further research

Research is needed to further assess KCD interventions in different contexts. I have applied the conceptual model in two cases in this thesis, but there are numerous other situations, in the water sector and outside, in which the presented new conceptual model could provide a reliable basis for structuring an assessment. Possible research topics are proposed in Table 8.2.

A deeper understanding of the influence of personality and attitude on job performance is required. The new KCD conceptual model goes some way to addressing and operationalising this issue, by means of looking at the personal/ethical competence as a subcomponent of competence, but it remains the mystery at the heart of knowledge and capacity development research.

Leadership has in this research been considered as part of the personal/ethical competence, but it could be argued that at the organisational or institutional level it is a component in itself, at the institutional level for example represented as policy entrepreneurs. In addition, in using the Multiple Streams Framework to explain the use of knowledge for policy making, I question whether the role of personal ambition receives sufficient attention in the framework as a driver for policy change.

The findings of this research may hold valid conclusions pertaining to the water sector in other developing and emerging economies. More research is suggested to further refine the analytical framework and investigative methods for data collection.

Table 8.2. A future research agenda for KCD

Research topic	Research Question (examples)
Empirical testing of the validity and completeness of the new conceptual model	What is the effect of personality and attitude on individual capacity development? To what extent can leadership be investigated as a personal/ethical competence and to what extent should be it considered a separate factor in itself at the organisational or institutional level? To what extent is personal ambition of policy entrepreneurs a driver for policy change? What would be an appropriate mix of competences for a public water management organisation such as the DGWR or the Rijkswaterstaat?
Applicability of the new KCD conceptual model to other case studies	How does the KCD model apply to sectors other than water resources management?
KCD dynamics	What effects of KCD interventions can be anticipated based on this conceptual model? What are the underlying causes and mechanisms determining the course of KCD interventions? How can KCD interventions be designed and adapted to increase their effectiveness?

8.6.2 Recommendations specifically for universities and other educational institutes

First, the framework cannot resolve decisions on the details of the required T-shaped competence profiles and skills mixes for specific water specialisations of individuals and organisations (e.g., on the relative importance of 'broad' competence versus 'in-depth' expertise) but it can help in outlining requirements that subsequently help guide graduate and post-graduate level water curricula improvements. It could be useful to design a number of modules in an educational programme that focus solely on the horizontal bar of the T-shaped competence profile. Students could then choose from a number of subjects that should not be part of their specialisation but should instead help them in connecting to other disciplines.

Third, it would be beneficial for the educational institute, the student and the participant's employer if there were more exchange about the needs of the home organisation and its ability to work with the new knowledge that an alumnus brings in after studying. The educational institute can take the employer's requirements into account and help the student to prepare for his or her return more thoroughly.

8.6.3 Recommendations for policy making in Overseas Development Assistance

In the second half of 2011 an online debate took place in The Netherlands on the policy for knowledge development in Overseas Development Assistance (ODA), in reaction to the letter to the House of Representatives by Mr. Ben Knapen, the State Secretary for International Development, in which he proposes a new policy for knowledge development (2011). The online debate has been summarised in a Special Report called

'Focus on the World' (Quak et al., 2011). First of all, the special report emphasises the need for multidisciplinary research to find solutions to complex global problems. I agree with this emphasis, as this research has also shown that to understand a knowledge system, it does not suffice to look at it from one discipline only. The three-level approach that I advocate with my KCD conceptual model automatically implies a multidisciplinary stance.

Secondly, the report acknowledges that the policy letter makes a cautious start in indicating that the traditional knowledge and development policy in which North and South are separated no longer corresponds with reality. This research shows that such a distinction is indeed untenable. The challenges and opportunities for the organisational and individual KCD are remarkably similar in both rich and poorer nations.

8.6.4   Recommendations for policy making in the DGWR and in the Rijkswaterstaat

To prepare the organisation for service delivery in an increasingly complex society, HRM could focus more on quality rather than quantity. HR quantities are pre-determined by the Ministry of State Administration, but quality can be influenced by developing a set of competence profiles that are needed in the organisation and applying these appropriately. A prerequisite is to have a clear idea about the organisation's identity and direction.

In both cases it is recommended to work on a strategy for succession planning, to harvest experiences before a staff member rotates to a new position or retires.

An evaluation of the outsourcing process to date would be advisable, to determine which competences should be in the organisation to be able to manage the outsourcing process.

8.7   FINAL CONSIDERATIONS

I started off with the overall aim to investigate the black box of KCD. Little research had been done before, making it hard to lean on the work of others but at the same time offering an unexplored exciting field of study. This research has provided a KCD conceptual model to assess the KCD system in a rigorous way; I established the nestedness of individuals in their organisation, which is in turn embedded in its institutional environment, resulting in a complicated web of interdependencies. The research provided evidence of the impact of international post-graduate education. This was not undertaken previously by means of academic research and provides useful material for educational institutions to evaluate their modus operandi. The research furthermore provides directions for shaping an organisational structure that is ready to cope with current and future water management challenges and in which the knowledge and capacities of water professionals can blossom. At the institutional level I indicate how culture and power relations shape the policy making process and organisational structure and how coalitions of professionals can influence this process. Suggestions for further application and improvement of the adapted KCD conceptual model were also made.

Research on KCD has proven to be a challenge. Because of its conceptual complexity, the continuously and unpredictably changing working environment of professional agencies and the need to observe any effects of KCD with very long delays, this subject will keep attracting controversy.

9 References

Nederland werkt met water: http://www.nederlandwerktmetwater.nl/ons_werk/, access: 23 May, 2012.

ADB: An introduction to results management - principles, implications and applications, ADB, Manila, 47, 2006.

Alaerts, G. J., and Herman, T.: Implementation Completion report. Water resources sector adjustment loan to Government of Indonesia, World Bank, 57, 2005.

Alaerts, G. J., and Dickinson, N. L.: Water for a changing world : developing local knowledge and capacity : proceedings of the International Symposium "Water for a Changing World - Developing Local Knowledge and Capacity", Delft, The Netherlands, June 13-15, 2007, International Symposium "Water for a Changing World - Developing Local, Knowledge and Capacity", Delft, The Netherlands, 2007, 2009.

Alaerts, G. J.: Knowledge and capacity development (KCD) as tool for institutional strengthening and change, in: Water for a changing world - Developing local knowledge and capacity, 1 ed., edited by: Alaerts, G. J., and Dickinson, N., Taylor & Francis Group, London, 22, 2009a.

Alaerts, G. J., and Kaspersma, J. M.: Progress and challenges in knowledge and capacity development, in: Capacity Development for improved water management, edited by: Blokland, M. W., Alaerts, G. J., Kaspersma, J. M., and Hare, M., Taylor and Francis, Delft, 327, 2009.

Albright, E. A.: Policy Change and Learning in Response to Extreme Flood Events in Hungary: An Advocacy Coalition Approach, Policy Studies Journal, 39, 485-511, 10.1111/j.1541-0072.2011.00418.x, 2011.

Appelgren, B., and Klohn, W.: Management of water scarcity: A focus on social capacities and options, Physics and Chemistry of the Earth, Part B, 24, 361-373, 10.1016/S1464-1909(99)00015-5, 1999.

Argyris, C., and Schön, D. A.: Organizational learning, Addison-Wesley Pub. Co., Reading, Mass., 1978.

Argyris, C.: On organizational learning, Blackwell Publishers, Cambridge, Mass., 1993.

Arnstein, S. R.: A Ladder Of Citizen Participation, Journal of the American Institute of Planners Journal of the American Institute of Planners, 35, 216-224, 1969.

Barney, J. B., and Ouchi, W. G.: Organizational economics, Jossey-Bass, San Francisco, 1986.

Baser, H., and Morgan, P.: Capacity, Change and Performance - study report, ECDPM, Maastricht, 166, 2008.

Baskerville, R. F.: Hofstede never studied culture, Accounting, Organizations and Society, 28, 1-14, 10.1016/s0361-3682(01)00048-4, 2003.

Bateson, D. S., Lalonde, A. B., Perron, L., and Senikas, V.: Methodology for Assessment and Development of Organization Capacity, JOGC -TORONTO-, 30, 888-895, 2008.

Belda, S., Boni, A., Peris, J., and Terol, L.: Rethinking capacity development for critica development practice. Inquiry into a postgraduate programme., Journal of International Development, 24, 571-584, 10.1002/jid.2850, 2012.

Bhagat, R. S., Kedia, B. L., Harveston, P. D., and Triandis, H. C.: Cultural Variations in the Cross-Border Transfer of Organizational Knowledge: An Integrative Framework, The Academy of Management Review, 27, 17, 2002.

Bhat, A., and Mollinga, P. P.: Transitions in Indonesian water policy: policy window through crisis, response through implementation, in: Water Policy Entrepreneurs: A Research Companion to Water Transitions around the globe, 1 ed., edited by: Huitema, D., and Meijerink, S., Edward Elgar Publishing Ltd, Cheltenham, 411, 2009.

Bird, J.: Indonesia in 1997: the tinderbox year, Asian Survey, 38, 168-176, 1998.

Biswas, A. K.: Capacity Building for Water Management: Some Personal Thoughts, International Journal of Water Resources Development, 12, 399-406, 1996.

Blankesteijn, M. L.: Tussen wetten en weten: de rol van kennis in waterbeheer in transitie, Boom Lemma, 2011.

Bogardi, J., and Hartvelt, F. J. A.: Towards a strategy on human capacity building for integrated water resources management and service delivery : water-education-training, UNESCO, Paris, 44 pp., 2002.

Bohlinger, S.: Competences as the core element of the European Qualifications Framework, European Journal for Vocational Training, 42/43, 17, 2007/2008.

Boonstra, J., and Muijen, J. v.: Leiderschap in organisaties : crisis in leiderschap - op zoek naar nieuwe wegen, Kluwer, Deventer, 2011.

Booth, A.: The evolving role of the central government in economic planning and policy making in Indonesia, Bulletin of Indonesian Economic Studies, 41, 197-219, 10.1080/00074910500117081, 2005.

Bosch, A., and Van der Ham, W.: Twee eeuwen Rijkswaterstaat 1798 - 1998, 2 ed., Europese Bibliotheek, Zaltbommel, 1998.

Bourget, P. G.: Key Lessons Learned from the Masters Degree Program in Water Resources Planning and Management, Journal of Contemporary Water Research & Education, 139, 55-57, 10.1111/j.1936-704X.2008.00021.x, 2008.

Bowman, J. S.: The success of failure: The paradox of performance pay, Review of Public Personnel Administration, 30, 70-88, 2010.

Brinkerhoff, D. W.: Organizational Legitimacy, Capacity, and Capacity Development, 8th Research Conference, Public Management Research Association, Los Angeles, sept 29 - Oct 1, 2005, 2005.

Brown, L., LaFond, A., and Macintyre, K.: Measuring capacity building, Carolina Population Center, University of North Carolina at Chapel Hill, Chapel Hill, NC, 2001.

Burns, T., and Stalker, G. M.: The management of innovation, Tavistock Publications, [London, 1961.

Cheetham, G., and Chivers, G.: Towards a holistic model of professional competence, Journal of European Industrial Training, 20, 11, 1996.

Cheetham, G., and Chivers, G.: Professions, competence and informal learning, Edward Elgar Publishing, Cheltenham UK, Northampton USA, 337 pp., 2005.

Child, J.: Organization Structure and Strategies of Control: A Replication of the Aston Study, Administrative Science Quarterly, 17, 163-177, 1972.

Chiva-Gómez, R.: The facilitating factors for organizational learning: bringing ideas from complex adaptive systems, Knowledge and Process Management, 10, 99-114, 2003.

Christensen, T.: Organization theory and the public sector : instrument, culture and myth, Routledge, London; New York, 191 pp., 2007.

Cleaver, F.: Reinventing Institutions: Bricolage and the Social Embeddedness of Natural Resource Management, The European Journal of Development Research, 14, 11-30, 2002.

Cleaver, F., Franks, T., Boesten, J., and Kiire, A.: Water governance and poverty - what works for the poor?, 2005.

Comim, F., Qizilbash, M., and Alkire, S.: The capability approach : concepts, measures and applications, Cambridge University Press, Cambridge, UK; New York, 2008.

Considine, M., Lewis, J. M., and Alexander, D.: Networks, innovation and public policy : politicians, bureaucrats and the pathways to change inside government, Palgrave Macmillan, Basingstoke [England]; New York, 2009.

Cosgrove, W. J., and Rijsberman, F. R.: World Water Vision: Making Water Everybody's Business, Earthscan, 2000.

Dalton, M.: Men who manage : fusions of feeling and theory in administration, Wiley, New York, 1959.

Datta, A., Jones, H., Febriany, V., Harris, D., Kumala Dewi, R., Wild, L., and Young, J.: The political economy of policy-making in Indonesia - Opportunities for improving the demand for use of knowledge, ODI, London, UK, 78, 2011.

Davenport, T. H., and Prusak, L.: Working knowledge : how organizations manage what they know, Harvard Business School Press, Boston, Mass, 1998.

Davies, R., and Dart, J.: The 'most significant change' (MSC) technique : a guide to its use, Rick Davies ; Jess Dart, Cambridge, UK; Chelsea, Australia, 2007.

de Haan, J., and Rotmans, J.: Patterns in transitions: Understanding complex chains of change, Technological Forecasting and Social Change, 78, 90-102, 10.1016/j.techfore.2010.10.008, 2011.

de Heer, J.: Strategie en verandering in organisaties onder druk : een onderzoek naar het besturen van herpositionering en transitieprocessen in organisaties van Verkeer en Waterstaat, VUGA, 's-Gravenhage, 1991.

Deaton, A.: Instruments, Randomization, and Learning about Development, Journal of Economic Literature, 48, 424-455, doi: 10.1257/jel.48.2.424, 2010.

Delamare Le Deist, F., and Winterton, J.: What is competence?, Human Resource Development International, 8, 27 - 46, 2005.

Deneulin, S., Nebel, M., and Sagovsky, N.: Transforming Unjust Structures - the capability approach, Springer, [New York], 2006.

Directoraat-Generaal Rijkswaterstaat: Anders omgaan met water : waterbeleid in de 21e eeuw, Ministerie van Verkeer en Waterstaat ; Postbus 51 Informatiedienst [distr.], Den Haag; [Den Haag], 2000.

Disco, C.: Remaking "Nature" The Ecological Turn in Dutch Water Management, Science, Technology & Human Values, 27, 206-235, 2002.

Domberger, S.: The contracting organization : a strategic guide to outsourcing, Oxford University Press, Oxford [England]; New York, 1998.

Downs, A.: Inside bureaucracy, Little, Brown, Boston, 1967.

Downs, T. J.: Making Sustainable Development Operational: Integrated Capacity Building for the Water Supply and Sanitation Sector in Mexico, Journal of Environmental Planning and Management, 44, 525-544, 2001.

Driscoll, D. L., Appiah-Yeboah, A., Salib, P., and Rupert, D. J.: Merging Qualitative and Quantitative Data in Mixed Methods Research: How To and Why Not, Ecological and Environmental Anthropology, 3, 10, 2007.

Duck, J.: Making the connection: improving virtual team performance through behavioral assessment profiling and behavioral cues, Developments in Business Simulation and Experiential Learning, 33, 2, 2006.

Outcome mapping building learning and reflection into development programs: http://site.ebrary.com/id/10137756, 2001.

Easterly, W.: Can the West Save Africa?, Journal of Economic Literature, 47, 373-447, doi: 10.1257/jel.47.2.373, 2009.

Edelenbos, J., van Buuren, A., and van Schie, N.: Co-producing knowledge: joint knowledge production between experts, bureaucrats and stakeholders in Dutch water management projects, Environmental Science & Policy, 14, 675-684, 10.1016/j.envsci.2011.04.004, 2011.

Cultural Factors as Co-Determinants of Participation in River Basin Management, 2007.

Eraut, M.: Informal learning in the workplace, Studies in Continuing Education, 26, 247-273, 10.1080/158037042000225245, 2004.

EuropeAid: Institutional Assessment and Capacity Development. Why, what and how?, European Commission, 31, 2005.

European Environmental Bureau: Letting the public have their say on water management - A snapshot analysis of Member States' consultations on water management issues and measures within the Water Framework Directive, European Environmental Bureau World Wide Fund for Nature, Brussels, 39, 2008.

FAO: Capacity development in irrigation and drainage.Issues, challenges and the way ahead, in: FAO Water Reports, International Commission on Irrigation and Drainage Fifty-fourth International Executive Council Meeting, Montpellier, France, 2004, 89,

Filmer, D., and Lindauer, D. L.: Does Indonesia have a 'low-pay' civil service?, Bulletin of Indonesian Economic Studies, 37, 17, 2001.

Fink, A.: How to sample in surveys, The survey kit, Sage Publications, Thousand Oaks, Cal. London, 1995.

Fiol, C. M., and Lyles, M. A.: Organizational Learning, The Academy of Management Review, 10, 803-813, 1985.

Forest, V.: Performance-related pay and work motivation: Theoretical and empirical perspectives for the French civil service, International Review of Administrative Sciences, 74, 325-339, 2008.

Fortuyn, P.: De puinhopen van acht jaar Paars : de wachtlijsten in de gezondheidszorg ... : een genadeloze analyse van de collectieve sector en aanbevelingen voor een krachtig herstelprogramma, Karakter ; Speakers Academy, Uithoorn; Rotterdam, 2002.

Fowler, A., and Ubels, J.: Multiple dimensions - The multi-faceted nature of capacity: two leading models, in: Capacity development in practice, 1 ed., edited by: Ubels, J., Acquaye-Baddoo, N.-A., and Fowler, A., Earthscan, London, 14, 2010.

Franks, T.: Capacity building and institutional development: reflections on water, Public Administration and Development, 19, 51-61, 1999.

Furlow, L.: Job Profiling: Building a Winning Team Using Behavioral Assessments, Journal of Nursing Administration, 30, 107-111, 2000.

Policy Initiatives Towards the Third Sector in International Perspective: http://dx.doi.org/10.1007/978-1-4419-1259-6, 2009.

Groves, R. M.: Survey methodology, J. Wiley, Hoboken, NJ, 2004.

Gupta, J., Termeer, C., Klostermann, J., Meijerink, S., van den Brink, M., Jong, P., Nooteboom, S., and Bergsma, E.: The Adaptive Capacity Wheel: A method to assess the inherent characteristics of institutions to enable the adaptive capacity of society, Environmental Science and Policy, 13, 459-471, 10.1016/j.envsci.2010.05.006, 2010.

Haas, P. M.: Introduction: epistemic communities and international policy coordination, International Organization, 46, 1-35, doi:10.1017/S0020818300001442, 1992.

Hamdy, A., Abu-Zeid, M., and Lacirignola, C.: Institutional Capacity Building for Water Sector Development, Water International, 23, 126-133, 1998.

Herman, T.: Implementation of Integrated Water Resources Management (IWRM) in Indonesia, World Bank, Jakarta, 52, 2007.

Hildebrand, M. E., and Grindle, M. S.: Building sustainable capacity : challenges for the public sector, Harvard Institute for International Development, Harvard Univerity, [Cambridge, Mass.], 1994.

Hofstede, G.: Cultures and organizations: software of the mind, McGraw-Hill, New York San Francisco, 1991.

Hofstede, G. H.: Culture's consequences : comparing values, behaviors, institutions, and organizations across nations, Sage Publications, Thousand Oaks, Calif., 2001.

Holbeche, L.: The high performance organization : creating dynamic stability and sustainable success, Elesevier, Amsterdam [u.a.], 2007.

House, R., Javidan, M., and Dorfman, P.: Project GLOBE: An Introduction, Applied Psychology, 50, 489-505, 10.1111/1464-0597.00070, 2001.

House, R. J., Global, L., and Organizational Behavior Effectiveness Research, P.: Culture, leadership, and organizations : the GLOBE study of 62 societies, Sage Publications, Thousand Oaks, Calif., 2004.

Houterman, J., Djoeachir, M., Susanto, R. H., and Steenbergen, F. v.: Indonesia: Water Resources Management in a Period of Transition and Reform - Country Case Study Report, WorldBank, Washington DC., 74, 2003.

Huisman, P.: Water in the Netherlands: managing checks and balances, NHV Special, Netherlands Hydrological Society, Utrecht, Delft, 2004.

Huitema, D., and Meijerink, S.: Policy dynamics in Dutch water management: analysing the contribution of policy entrepreneurs to policy change, in: Water Policy Entrepreneurs - A Research Companion to Water Transitions around the Globe, edited by: Huitema, D., and Meijerink, S., IWA Publising, Northampton, MA, USA, 19, 2009a.

Inglehart, R., and Baker, W. E.: Modernization, cultural change, and the persistence of traditional values, American Sociological Review, 65, 19-51, 2000.

Human beliefs and values a cross-cultural sourcebook based on the 1999-2002 values surveys: http://www.netlibrary.com/urlapi.asp?action=summary&v=1&bookid=107501, 2004.

Inglehart, R., and Welzel, C.: Modernization, cultural change, and democracy : the human development sequence, Cambridge University Press, Cambridge, UK; New York, 2005.

De exodus van ambtenaren: ervaren rotten met prepensioen: http://www.inoverheid.nl/artikel/nieuws/1112074/de-exodus-van-ambtenaren-ervaren-rotten-met-prepensioen.html, access: 3 June 2012, 2006.

Jones, H.: Policy-making as a discourse: a review of recent knowledge-to-policy literature, IKM Emergent Research Programme, European Association of Development Research and Training Institutes, Bonn, Germany, 2009.

Jones, M.: Hofstede - Culturally questionable?, Oxford Business & Economics Conference, Oxford, UK, 24-26 June 2007, 2007.

Kainz, H. P.: G.W.F. Hegel : the philosophical system, Twayne Publishers, : Prentice Hall International, New York; London, 1996.

Karstens, S., De Quelerij, L., Van Wijngaarden, M., Voogt, L., Maccabiani, J., Van de Giesen , N., Bremmer, C., Van den Bosch, R., and Bergme, P.: Innovatiecontract Deltatechnologie 2.0, 45, 2011.

Keijzer, N., Spierings, E., Phlix, G., and Fowler, A.: Bringing the invisible into perspective. Reference paper for using the 5Cs framework to plan, monitor and evaluate capacity and results of capacity development processes, ECDPM, Maastricht, 61, 2011.

Kim, D. H.: The link between individual and organizational learning, in: The strategic management of intellectual capital, edited by: Klein, D. A., Resources for the knowledge based economy, Butterworth-Heinemann, Boston, 1993.

Kingdon, J. W.: Agendas, alternatives, and public policies, Little, Brown, Boston, 1984.

Kingdon, J. W.: Agendas, alternatives, and public policies, Longman, New York, 1995.

Knafo, A., Roccas, S., and Sagiv, L.: The Value of Values in Cross-Cultural Research: A Special Issue in Honor of Shalom Schwartz, Journal of Cross-Cultural Psychology, 42, 178-185, 10.1177/0022022110396863, 2011.

Kolb, D. A.: Experiential learning : experience as the source of learning and development, Prentice-Hall, Englewood Cliffs, N.J., 1984.

Kuhn, T. S.: The structure of scientific revolutions, the University of Chicago Press, Chicago, 1962.

Kuklys, W.: Amartya Sen's capability approach, Springer, Berlin; New York, 2005.

Land, T., Hauck, V., and Baser, H.: Capacity development: between planned interventions and emergent processes. Implications for development cooperation., ECDPM, Maastricht, 8, 2009.

Lave, J., and Wenger, E.: Situated learning : legitimate peripheral participation, Cambridge University Press, Cambridge [England]; New York, 1991.

Levitt, B., and March, J. G.: Organizational Learning, Annual Review of Sociology, 14, 319-340, 1988.

Liao, S. H., Fei, W. C., and Liu, C. T.: Relationships between knowledge inertia, organizational learning and organization innovation, Technovation, 28, 183-195, DOI 10.1016/j.technovation.2007.11.005, 2008.

Lintsen, H.: Two Centuries of Central Water Management in the Netherlands, Technology and Culture, 43, 549-568, 2002.

Lintsen, H., Disco, N., and Geels, F.: Hoe innovatief is de Rijkswaterstaat? Een langetermijn-analyse van de wegen, rivieren en kustverdediging (1950-2000), Tijdschrift voor de waterstaatsgeschiedenis, 13, 15, 2004.

Litfin, K. T.: Ozone discourses : science and politics in global environmental cooperation, Columbia Univ. Press, New York, 1994.

Loosveldt, G.: Face-to-face interviews, in: International handbook of survey methodology, edited by: De Leeuw, E. D., Hox, J. J., and Dillman, D. A., Taylor and Francis, New York
Abingdon, 549, 2008.

Loucks, D. P.: Educating Future Water Resources Managers, Journal of Contemporary Water Research & Education, 139, 17-22, 10.1111/j.1936-704X.2008.00014.x, 2008.

Lusthaus, C., Adrien, M. H., and Perstinger, M.: Capacity Development: Definitions, Issues and Implications for Planning, Monitoring and Evaluation, Universalia Occasional Paper, 35, 21, 1999.

Lusthaus, C., Adrien, M.-H., Anderson, G., Carden, F., and Montalvan, G. P.: Organizational Assessment. A Framework for Improving Performance, IDB-IDRC, Washington, Ottawa, 2002.

Luzi, S., Abdelmoghny Hamouda, M., Sigrist, F., and Tauchnitz, E.: Water Policy Networks in Egypt and Ethiopia, The Journal of Environment & Development, 17, 238-268, 10.1177/1070496508320205, 2008.

Madani, K.: Game theory and water resources, Journal of Hydrology, 381, 225-238, 10.1016/j.jhydrol.2009.11.045, 2010.

Manfreda, K. L., and Vehovar, V.: Internet surveys, in: International handbook of survey methodology, edited by: De Leeuw, E. D., Hox, J. J., and Dillman, D. A., Taylor and Francis, New York
Abingdon, 549, 2008.

Marsick, V. J., and Watkins, K. E.: Informal and Incidental Learning, New Directions for Adult and Continuing Education, 2001, 25-34, 10.1002/ace.5, 2001.

Marsick, V. J., and Watkins, A. E.: Demonstrating the Value of an Organization's Learning Culture: The Dimensions of the Learning Organization Questionnaire, Advances in Developing Human Resources, 5, 19, 10.1177/1523422303251341, 2003.

Marston, W. M.: Emotions of normal people, K. Paul, Trench, Trubner & Co. ltd.; Harcourt, Brace and Company, London; New York, 1928.

McClelland, D.: A guide to job competency assessment, McBer, Boston, MA, 1976.

McClelland, D.: Identifying Competencies with Behavioral-Event Interviews, Psychological Science, 9, 331-339, 10.1111/1467-9280.00065, 1998.

McDaniel Jr., R. R.: Management Strategies for Complex Adaptive Systems Sensemaking, Learning, and Improvisation, Performance Improvement Quarterly, 20, 21-41, 2007.

Mcleod, R. H.: The struggle to regain effective government under democracy in Indonesia, Bulletin of Indonesian Economic Studies, 41, 367 - 386, 2005.

McSweeney, B.: The Essentials of Scholarship: A Reply to Hofstede, Human Relations, 55, 9, 2002a.

McSweeney, B.: Hofstede's Identification of National Cultural Differences – A Triumph of Faith a Failure of Analysis, Human Relations, 55, 29, 2002b.

Meijerink, S., and Huitema, D.: Understanding and managing water policy transitions: a policy science perspective, in: Water policy entrepreneurs: a research companion to water transitions around the globe, edited by: Meijerink, S., and Huitema, D., Edward Elgar Publishing Ltd, Cheltenham, 411, 2009.

Metze, M.: Veranderend getij. Rijkswaterstaat in crisis : de bouwfraude, de Heeren Zeventien, een nieuwe man, de coup, het verzet, de 'generaal', de managersrevolutie, een nieuwe koers. Het verhaal van binnenuit, Balans, Amsterdam, 2008.

Ministerie van Verkeer en Waterstaat: Living with water : towards an integral water policy, Ministry of Transport and Publics Works], The Hague, 1986.

Ministry of Public Works - Directorate General of Water Resources: Development of a Master Plan for Capacity Development for the Ministry of Public Works – Directorate General for Water Resources, Ministry of Public Works, Jakarta, 63, 2010.

Minkov, M.: What makes us different and similar : a new interpretation of the World Values Survey and other cross-cultural data, Klasika i Stil Publishing House, Sofia, 2007.

Minkov, M., and Hofstede, G.: The evolution of Hofstede's doctrine, Cross Cultural Management, 18, 10-20, 2011.

Mintzberg, H.: The structuring of organizations, Prentice-Hall, Englewood Cliffs, N.J., 1979.

Mintzberg, H.: Structure in 5's: A Synthesis of the Research on Organization Design, Management Science, 26, 322-341, 1980.

Mintzberg, H.: Structure in fives : designing effective organizations, Prentice-Hall, Englewood Cliffs, N.J., 1983.

Mizrahi, Y.: Capacity Enhancement Indicators: review of the literature, World Bank Institute, Washington DC., 38, 2004.

Mollinga, P. P.: Towards the transdisciplinary engineer: Incorporating ecology, equity and democracy concerns into water professionals' attitudes, skills and knowledge, Irrigation and Drainage, 58, S195-S204, 2009.

Morgan, P.: Capacity Building: an overview, Workshop on Capacity Development, Ottawa, november 22-23, 1993.

Morgan, P.: The Concept of Capacity, report, 19, 2006.

Morgan, P. J., Baser, H., and Morin, D.: Developing capacity for managing public service reform: The Tanzania experience 2000-2008, Public administration and development, 30, 27-37, 2010.

Mugisha, S., and Brown, A.: Patience and action pays: A comparative analysis of WSS reforms in three East African cities, Water Policy, 12, 654-674, 2010.

Muizer, A. P., and Leusink, A.: Economische betekenis van de waterzuiveringstechnologie, EIM, Zoetermeer, 101, 2005.

Muizer, A. P., Morselt, T. T., Verhoeven, W. H. J., and Folkeringa, M.: Het Nederlandse deltatechnologie-cluster - Economische waarde, internationale concurrentiekracht en arbeidsmarktperspectieven, Panteia, Zoetermeer, 196, 2010.

Nash, J. E., Eagleson, P. S., Philip, J. R., Van Der Molen, W. H., and Klemeš, V.: The education of hydrologists (Report of an IAHS/UNESCO Panel on hydrological education), Hydrological Sciences Journal, 35, 597-607, 10.1080/02626669009492466, 1990.

Netherlands Water Partnership: Water 2020 - Wereldleiders in water, Netherlands Water Partnership, Den Haag, 52, 2011.

Nonaka, I.: A Dynamic Theory of Organizational Knowledge Creation, Organization Science, 5, 14-37, 1994.

Nonaka, I., and Takeuchi, H.: The Knowledge-Creating Company - How Japanese Companies Create the Dynamics of Innovation, 1 ed., Oxford University Press, Inc., New York, 284 pp., 1995.

Nonaka, I., and Takeuchi, H.: The wise leader, Harvard business review, 89, 58-67, 146, 2011.

North, D. C.: Institutions, Institutional Change and Economic Performance, Cambridge University Press, 152 pp., 1990.

OECD: Paris declaration on aid effectiveness -Ownership, Harmonisation, Alignment, Results and Mutual Accountability, 2005, 12,

OECD: The Challenge of Capacity Development - Working Towards Good Practice, DAC Guidelines and Reference Series, 45, 2006.

OECD: Paris Declaration on Aid Effectiveness and the Accra Agenda for Action, OECD, Paris, 166, 2008.

Oskam, I. F.: T-shaped engineers for interdisciplinary innovation: an attractive perspective for young people as well as a must for innovative organisations, 37th Annual Conference - Attracting students in Engineering, Rotterdam, The Netherlands, 1 - 4 July 2009, 2009.

Ostrom, E.: Crafting institutions for self-governing irrigation systems, ICS Press ; Distributed to the trade by National Book Network, San Francisco, Calif.; Lanham, Md., 1992.

Ostrom, E.: Understanding institutional diversity, Princeton University Press, Princeton, 2005.

Oswald, K.: Reflecting collectively on capacities for change, Blackwell Publishing, Oxford, 2010.

Otjes, S.: The Fortuyn effect revisited: How did the LPF affect the Dutch parliamentary party system?, Acta Politica, 46, 400-424, 2011.

Otoo, S., Agapitova, N., and Behrens, J.: Capacity Development Results Framework, World Bank Washington DC, 100, 2009.

Pahl-Wostl, C.: Towards sustainability in the water sector - The importance of human actors and processes of social learning, Aquatic Sciences, 64, 17, 2002.

Pahl-Wostl, C., Craps, M., Dewulf, A., Mostert, E., Tabara, D., and Taillieu, T.: Social Learning and Water Resources Management, Ecology and Society, 12, 19, 2007.

Pascual Sanz, M., Schouten, M., Alaerts, G., and Van Tulder, R.: Appraising water operator's capacity to reduce non revenue water. The case of Bantyre water board, 2011.

Pascual Sanz, M.: Peer-partnering for organizational development of water operators PhD, Erasmus University, Rotterdam, Rotterdam, forthcoming.

Pavlov, I.: Conditioned Reflexes: An Investigation of the Physiological Activity of the Cerebral Cortex, Oxford University Press, London, 1927.

Pearson, J.: LenCD Learning Package on Capacity Development - Part I: The Core Concept, Part II: How to..., Part III: Trainer-facilitator's guide, Learning Network on Capacity Development, 92, 2011.

Polanyi, M.: The Tacit Dimension, Doubleday & company, New York, 108 pp., 1966.

Polidano, C.: Measuring Public Sector Capacity, World Development, 28, 805-822, 2000.

Prahalad, C. K., and Bettis, R. A.: The Dominant Logic: A New Linkage between Diversity and Performance, Strategic Management Journal, 7, 485-501, 1986.

Prahalad, C. K., and Hamel, G.: The core competence of the corporation, Harvard Business School Pub. Corp., [Boston, MA], 1993.

Rainey, H. G.: Understanding and managing public organizations, 3 ed., The Jossey-Bass nonprofit and public management series, Jossey-Bass, San Francisco, 2003.

Ramalingam, B., Jones, H., Reba, T., and Young, J.: Exploring the science of complexity - Ideas and implications for development and humanitarian efforts, Overseas Development Institute, London, 71, 2008.

Ramu, K. V.: Brantas River Basin Case Study, World Bank, Washington, 76, 2004.

Rand Corporation, Rijkswaterstaat, and Waterloopkundig Laboratorium: Policy analysis of water management for the Netherlands, Rand, Santa Monica, CA, 1981.

Rhodes, J., Walsh, P., and Lok, P.: Convergence and divergence issues in strategic management - Indonesia's experience with the Balanced Scorecard in HR management, International Journal of Human Resource Management, 19, 1170-1185, 2008.

Rijkswaterstaat Organisation: http://www.rijkswaterstaat.nl/en/about_us/, access: 1 June, 2012.

Robeyns, I.: An unworkable idea or a promising alternative? : Sen's capability approach re-examined, Katholieke Universiteit Leuven. , Leuven, 2000.

Robeyns, I.: The Capability Approach: a theoretical survey, Journal of Human Development, 6, 93-117, 10.1080/146498805200034266, 2005.

Rohdewohld, R.: Public Administration in Indonesia, Montech Pty Ltd., 1995.

Rooijendijk, C.: Waterwolven : een geschiedenis van stormvloeden, dijkenbouwers en droogmakers, Atlas, Amsterdam [etc.], 2009.

Sabatier, P. A., and Jenkins-Smith, H. C.: Policy change and learning : an advocacy coalition approach, Westview Press, Boulder, Co., 1993.

Sabatier, P. A.: Theories of the policy process, Westview Press, Boulder, Colo., 2007.

Sachs, J.: The end of poverty : economic possibilities for our time, Penguin Press, New York, 2005.

Saldaña, J.: The coding manual for qualitative researchers, Sage, London; Thousand Oaks, Calif., 2009.

Sarsito, T.: Javanese culture as the source of legitimacy for Soeharto's government, Asia Europe Journal, 4, 447-461, 2006.

Schein, E. H.: Organizational culture and leadership, Jossey-Bass Publishers, San Francisco, 1985.

Schein, E. H.: Three cultures of management: the key to organizational learning, Sloan management review., 38, 9, 1996.

Schmidt, V. A., and Radaelli, C. M.: Policy Change and Discourse in Europe: Conceptual and Methodological Issues, West European Politics, 27, 183-210, 10.1080/0140238042000214874, 2004.

Schwartz, K.: Towards a masterplan for capacity building in the Indonesian water resources sector - project findings and recommendations, UNESCO-IHE, Delft, 40, 2008.

Schwartz, S. H.: Universals in the Content and Structure of Values: Theoretical Advances and Empirical Tests in 20 Countries, in: Advances in Experimental Social Psychology, edited by: Mark, P. Z., Academic Press, 1-65, 1992.

Schwartz, S. H., and Sagiv, L.: Identifying Culture-Specifics in the Content and Structure of Values, Journal of Cross-Cultural Psychology, 26, 92-116, 10.1177/0022022195261007, 1995.

Sen, A.: Development as freedom, Oxford University Press, Oxford, 366 pp., 1999.

Senge, P. M.: The fifth discipline: the art and practice of the learning organization, Doubleday/Currency, New York, 1990.

Sherlock, S.: Knowledge for policy: Regulatory Obstacles to the growth of a knowledge market in Indonesia, AusAID, 45, 2010.

Slinger, J. H., Hilders, M., and Juizo, D.: The practice of transboundary decision-making on the Incomati River: elucidating underlying factors and their implications for institutional design, Ecology and Society, 15, 2010.

Snell, R., and Hong, J.: Organizational Learning in Asia, in: Handbook of organizational learning and knowledge management, 2 ed., edited by: Easterby-Smith, M., and Lyles, M. A., Wiley, Chichester, UK, 635 - 658 2011.

Suhardiman, D.: Bureaucratic Designs - The Paradox of Irrigation Management Transfer in Indonesia, PhD, Irrigation and Water Engineering Group, Wageningen University, Wageningen, 285 pp., 2008.

Sultana, R.: Competence and competence frameworks in career guidance: complex and contested concepts, International Journal for Educational and Vocational Guidance, 9, 15-30, 2009.

Sveiby, K.-E.: A Knowledge-based Theory of the Firm to guide Strategy Formulation, Journal of Intellectual Capital, 2, 15, 2001a.

Synnerstrom, S.: The civil service: towards efficiency, effectiveness and honesty, in: Indonesia : Democracy and the promise of good governance, edited by: Mcleod, R. H., and MacIntyre, A. J., Institute of Southeast Asian Studies, Canberra, Australia, 19, 2007.

Teasley, R. L., and McKinney, D. C.: Evaluating water resource management in the transboundary Rio Grande/Bravo using cooperative game theory, 2010, 2194-2203,

Therkildsen, O.: Public sector reform in a poor, aid-dependent country, Tanzania, Public Administration and Development, 20, 61-71, 10.1002/1099-162X(200002)20:1<61::AID-PAD101>3.0.CO;2-T, 2000.

Tijssen, R., Nederhof, A., Van Leeuwen, T., Hollanders, H., Kanerva, M., and Van den Berg, P.: Wetenschaps- en Technologie-indicatoren Nederlands Observatorium van Wetenschap en Technologie (NOWT), 176, 2010.

Tjakraatmadja, J. H., Martini, L., and Wicaksono, A.: Knowledge Sharing in Indonesian Context: Institut Teknologi Bandung (ITB) as Potential Knowledge Hub to Create Value from Academia, Business and Government Linkages, 4th International Research Conference on Asian Business: "Knowledge Architecture for Development: Challenges Ahead for Asian Business and Governance", Singapore, 24-25 March 2008, 2011.

Toussaint, B.: Waterbeheer, Centraal Bureau voor Genealogie, Den Haag, 2003.

Tropp, H.: Water governance: trends and needs for new capacity development, Water Policy, 9, 19-30, 2007.

Tsoukas, H.: Do we really understand tacit knowledge?, Knowledge Economy and Society Seminar, London, 14 June, 2002.

Ubels, J.: Capacity development in practice, Earthscan, London, 2010.

Uhlenbrook, S., and De Jong, E.: T-shaped competency profile for water professionals of the future, Hydrology and Earth System Sciences Discussions, 9, 22, 2012.

UNDP: Capacity Development, UNDP New York, 42, 1997.

Human Development Report: http://hdr.undp.org/en/humandev/, access: 20 June, 2008b.

UNDP: Measuring capacity, UNDP, New York, 40, 2010.

Usman, S.: Indonesia's Decentralization Policy: Initial Experiences and Emerging Problems, Third EUROSEAS Conference Panel on Decentralization and Democratization in Southeast Asia., London, September, 2001.

Van de Port, P., and Veenswijk, M.: Kennisbehoefte in Kleuren - Rijkswaterstaat in een veranderende Context, Vrije Universiteit, Amsterdam, 2006.

Van de Ven, G. P.: Man-made lowlands: history of water management and land reclamation in the Netherlands, 4th revised ed., Matrijs, Utrecht, 2004.

Van den Brink, M. A.: Rijkswaterstaat on the horns of a dilemma, PhD, Management Sciences, Radboud Universiteit Nijmegen, Delft, 335 pp., 2009.

van der Brugge, R., Rotmans, J., and Loorbach, D.: The transition in Dutch water management, Regional Environmental Change, 5, 164-176, 2005.

Van der Brugge, R.: Transition dynamics in social-ecological systems: The case of Dutch water management, PhD, Dutch Research Institute For Transitions (DRIFT), Faculty of Social Sciences, Erasmus University, Rotterdam, 264 pp., 2009.

Van der Ham, W.: Heersen en beheersen : Rijkswaterstaat in de twintigste eeuw, Europese Bibliotheek, Zaltbommel, 1999.

van Heezik, A.: Strijd om de rivieren : 200 jaar rivierenbeleid in Nederland of de opkomst en ondergang van het streven naar de normale rivier, Beleidsresearch.nl | Van Heezik Beleidsresearch ; Directoraat-Generaal Rijkswaterstaat/Ministerie van Verkeer en Waterstaat, Haarlem; Den Haag, 2008.

van Overveld, P. J. M., Hermans, L. M., and Verliefde, A. R. D.: The use of technical knowledge in European water policy-making, Environmental Policy and Governance, 20, 322-335, 10.1002/eet.546, 2010.

Vernon, T., and Werner, B.: Authentic innovation: The role of apprentice learning in engineering education, in: 2009 ASME International Mechanical Engineering Congress and Exposition, IMECE2009, Lake Buena Vista, FL, 2010, 439-447,

Vreugdenhil, H., Slinger, J. H., Thissen, W., and Ker Rault, P.: Pilot projects in water management, Ecology and Society, 15, 13, 2010.

Wade, R., and Chambers, R.: Managing the main system: canal irrigation's blind spot, Economic and political weekly, 15, 5, 1980.

Wagener, T., Weiler, M., McGlynn, B., Gooseff, M., Meixner, T., Marshall, L., McGuire, K., and McHale, M.: Taking the pulse of hydrology education, Hydrological Processes, 21, 1789-1792, 10.1002/hyp.6766, 2007.

Wagener, T., Sivapalan, M., Troch, P. A., McGlynn, B. L., Harman, C. J., Gupta, H. V., Kumar, P., Rao, P. S. C., Basu, N. B., and Wilson, J. S.: The future of hydrology: An evolving science for a changing world, Water Resources Research, 46, 2010.

Weggeman, M.: Kennismanagement: inrichting en besturing van kennisintensieve organisaties, Scriptum management, Scriptum, Schiedam, 1997.

Weible, C. M., Sabatier, P. A., Jenkins-Smith, H. C., Nohrstedt, D., Henry, A. D., and deLeon, P.: A Quarter Century of the Advocacy Coalition Framework: An Introduction to the Special Issue, Policy Studies Journal, 39, 349-360, 10.1111/j.1541-0072.2011.00412.x, 2011.

Weiss, C. H.: The Interface between Evaluation and Public Policy, Evaluation, 5, 468-486, 10.1177/135638909900500408, 1999.

Westwood, R.: Harmony and Patriarchy: The Cultural Basis for 'Paternalistic Headship' Among the Overseas Chinese, Organization Studies, 18, 445-480, 10.1177/017084069701800305, 1997.

White, R. W.: Motivation reconsidered: the concept of competence, Psychological Review, 66, 36, 1959.

Whyte, A.: Landscape Analysis of Donor Trends in International Development, The Rockefeller Foundation, New York, 84, 2004.

Woodhill, J.: Capacities for Institutional Innovation: A Complexity Perspective, IDS Bulletin, 41, 47-59, 10.1111/j.1759-5436.2010.00136.x, 2010.

World Bank - IEG: Using Training to Build Capacity for Development - An Evaluation of the World Bank's Project-Based and WBI Training, WorldBank IEG, Washington DC., 119, 2008.

World Water Assessment Programme: The United Nations World Water Development Report 4: Managing Water under Uncertainty and Risk (Vol. 1), Knowledge Base (Vol. 2) and Facing the Challenges (Vol. 3), Unesco, Paris, 407, 2012.

Yin, R. K.: Case Study Research, 3 ed., Applied Social Research Methods Series, 5, edited by: Bickman, L., and Rog, D. J., SAGE Publications, Thousand Oaks, 181 pp., 2003.

Yusoff, M. S. B. M.: The public service as a learning organization: The Malaysian experience, International Review of Administrative Sciences, 71, 463-474, 2005.

Zevenbergen, A., and Boer, J.: De deskundige: leerling en leermeester : een halve eeuw uitzending van ontwikkelingswerkers, Ministerie van Buitenlandse Zaken, Den Haag, 2002.

Zinke, J.: Monitoring and Evaluation of capacity and capacity development, ECDPM, Maastricht, 38, 2006.

10 Nederlandse Samenvatting

Sinds de jaren vijftig is international ontwikkelingssamenwerking gestaag gegroeid. De afgelopen zestig jaar kan gezien worden als een lange-termijn proces waarin men heeft geleerd wat ontwikkeling eigenlijk omvat (Zevenbergen and Boer, 2002). De doelmatigheid van ontwikkelingssamenwerking is een onderwerp geworden in het publieke debat. Onder niet-economen gaat men er vaak vanuit dat als de fondsen beschikbaar waren, armoede zou verdwijnen, en ook sommige economen, zoals Sachs (2005) zijn het daarmee eens (Deaton, 2010). Anderen, zoals Easterly (2009), geloven dat een 'bottom-up' aanpak en niet noodzakelijkerwijs de beschikbaarheid van enorme bedragen meer succesvol zou zijn, omdat deze aanpak een stem geeft aan lokale gemeenschappen om zelf hun behoeften aan te geven. Meer en meer gedocumenteerde ervaringen geven aan dat geld alleen niet de oplossing is, en dat capaciteit en kennis in toenemende mate gezien moeten worden als sleutelbeperkingen voor goede besluitvorming, voor het absorberen van fondsen, en voor effectieve resultaten. Volgens de verklaring over de effectiviteit van hulp die is opgesteld in Parijs (OECD, 2005), zullen ontwikkelingsdoelen in veel van de armste landen niet gehaald worden, zelfs als de financiering structureel stijgt, als de ontwikkeling van duurzame capaciteit niet meer aandacht krijgt.

Vanwege haar complexiteit is de water sector in het bijzonder afhankelijk van effectieve instituties, en dus van capaciteit en een solide kennisbasis op individueel en institutioneel niveau (Cosgrove and Rijsberman, 2000). Het is daarom niet verrassend dat de water sector een van eerste sectoren was die initiatieven op gebied van kennis- en capaciteitsontwikkeling (KCD) ontplooide (Alaerts et al., 1991; Hamdy et al., 1998; Appelgren and Klohn, 1999; Downs, 2001; Bogardi and Hartvelt, 2002; FAO, 2004; Alaerts, 2009b; Whyte, 2004). Desalniettemin blijven het beheer van watersystemen en het leveren van water aan burgers, landbouw en industrie enorme uitdagingen.

Het doel van dit proefschrift is om het dynamische proces van kennis- en capaciteitsontwikkeling samen met de vele contextuele factoren die capaciteit van het individuele niveau tot aan het systeemniveau beter te begrijpen, en zo de effectiviteit van kennis- en capaciteitsontwikkelingsprogramma's en interventies te verbeteren.

De verwerkte literatuur laat een spanning zien tussen KCD modellen die werkbaar en operationeel zijn, en modellen die meer complex zijn, en daarmee een betere weerspiegeling kunnen zijn van de werkelijkheid, maar moeilijker toepasbaar in de praktijk. Het aangepaste KCD conceptuele model dat ik gebruik (Alaerts and Kaspersma, 2009) is volledig in de zin dat het KCD als geheel beschouwt op het individuele, organisatie en institutionele niveau. Het geeft verder aandacht aan de interactie tussen de drie niveaus. Andere modellen houden rekening met de invloed van de andere niveaus naast het primaire niveau van analyse, maar beoordelen niet de bestaande kennis en capaciteit op alle drie niveaus. Ze richten hun aandacht doorgaans op slechts één niveau. Het aangepaste KCD model dient als basis, als structureringskader in het onderzoek naar KCD in de water sector. Daarnaast gebruik ik theorieën uit aanpalende velden zoals 'human resource development' (HRD), didactiek, organisatie- en managementtheorie en beleidsanalyse om de relaties tussen de verschillende componenten van het KCD systeem te verklaren (Hoofdstuk 2). Op individueel niveau gebruik ik theorieën over professionele competenties (Cheetham and Chivers, 2005; Sultana, 2009; Oskam, 2009) om de samenstelling van kennis en capaciteiten en de combinatie van verschillende competenties die waterdeskundigen nodig kunnen hebben uit te leggen (Hoofdstuk 6). Op het niveau van organisaties gebruik ik Burns and Stalker's (1961) indeling in mechanistische en meer organische organisatiestructuren om uit te leggen hoe de formele organisatiestructuur KCD beïnvloedt (Hoofdstuk 5). Op het institutionele niveau put ik uit theorie over coalitievorming (Sabatier and Jenkins-Smith, 1993) en een theoretisch kader voor beleidsanalyse, het 'meer-stromen kader' (MSF), ontwikkeld door Kingdon (1995). Met deze twee kaders verklaar ik hoe coalities voortdurend hun agenda voor het voetlicht moeten brengen en druk moeten uitoefenen, om bestaande beleidsregimes te beïnvloeden. De agenda van die coalities bevat vaak nieuwe kennis en capaciteiten. Vaak zijn er externe gebeurtenissen nodig (de probleem 'stroom') die een politieke reactie uitlokken (de politieke 'stroom'), om het bestaande beleid (de beleids-'stroom') te veranderen, en over te gaan naar een nieuw paradigma waarin de nieuwe kennis en capaciteiten een plek hebben (Hoofdstuk 4).

Ik gebruik het aangepaste kader voor KCD samen met de aanvullende theorieën om KCD te bestuderen in twee overheidsorganisaties die ik representatief acht voor de water sector in de wereld, over een langere periode, en in hun institutionele context. De eerste casus is het Directoraat Generaal Water van het Ministerie voor Openbare Werken in Indonesië. In deze casus neem ik internationaal postgraduaats onderwijs als startpunt. Mijn hypothese is dat internationaal postgraduaats onderwijs relatief belangrijk is in een maatschappij waar weinig andere KCD mechanismen beschikbaar lijken te zijn. In veel ontwikkelingslanden en landen in transitie is internationaal postgraduaats onderwijs belangrijk om toegang te krijgen tot kennis die lokaal niet beschikbaar is. De tweede casus is de uitvoeringsorganisatie van het Nederlandse Ministerie van Infrastructuur en Milieu, Rijkswaterstaat. Deze organisatie werd geselecteerd om te kunnen onderzoeken hoe kennis en capaciteit ontwikkeld worden en besluitvorming beïnvloeden in een gelijksoortige organisatie als in de eerste casus, maar in een relatief welvarende economie. Mijn aanname is dat er een grotere variëteit aan KCD mechanismen beschikbaar is om nieuwe kennis en capaciteiten te ontwikkelen in de tweede casus.

Om te analyseren hoe waterdeskundigen nieuwe kennis en capaciteiten verkrijgen, heb ik gebruik gemaakt van questionnaires en semi-gestructureerde interviews. Daarnaast heb ik een historische analyse van zowel de Nederlandse als Indonesische watersector gemaakt om te begrijpen hoe de culturele en omgevingsfactoren en -prioriteiten in de maatschappij het gebruik van KCD mechanismen hebben beïnvloed door de tijd heen. In de historische analyse heb ik drie verschillende fasen geïdentificeerd, zoals getoond in Tabel 10.1, die gekenmerkt worden door soortgelijke paradigma's. De fasen beslaan een periode van ongeveer 40 jaar. De introductie van deze fasen is nodig om de analyse mogelijk te maken van eenzelfde systeem onder verschillende omstandigheden, en met verschillende institutionele en contextuele parameters.

De methodologische differentiatie van respondenten in de Indonesische casus als functie van lokaal of internationaal postgraduaats onderwijs en in beide casus de differentiatie van de administratieve en politieke contexten in de landen en sectoren per fase, was zinvol voor het verkrijgen van gedetailleerde inzichten in de ontwikkeling van competenties in beide sectoren. In beide casus zoek ik naar parallellen en verschillen om algemene regels af te leiden voor KCD processen.

Ten eerste, op het institutionele niveau, liggen de verschillen tussen de casus in het politieke draagvlak en de manier waarop de coalities erin slagen om druk uit te oefenen en hun agenda voor het voetlicht te brengen. In de Indonesische casus werd de kennis die door de gemeenschap van donors, internationale adviseurs en hervormingsgezinde ambtenaren werd binnengebracht als bedreigend ervaren voor de status quo in het Directoraat Generaal, net zoals de Rijkswaterstaat zich bedreigd voelde door de milieulobby in de jaren 70 (Fase II). Echter, in de Indonesische casus slaagde de coalitie er niet in om het bestaande waterbeleid van Fase I te beïnvloeden, omdat er geen draagvlak was voor integraal water beheer en loyaliteit aan het regime belangrijker bleek. In Indonesië leverde de financiële en politieke crisis in 1998 de opening en het momentum om naar een nieuw paradigma te veranderen. Overeenkomsten met de Nederlandse casus in Fase III worden gevonden in de toegenomen aandacht voor verantwoording en transparantie, en het moeizame beheer van het vermarktingsproces, vanwege het gebrek aan technische competentie in beide organisaties.

Table 10.1 Algemene beschrijving van transities en paradigmatische fasen in de Indonesische en Nederlandse casus.

Indonesische casus	Nederlandse casus
Fase I (1970 - 1987) Ontwikkeling van infrastructuur Technocratisch Weinig ruimte voor andere vakgebieden dan techniek Organisatie gebaseerd op senioriteit	Fase I (1950 – 1970) Ontwikkeling van infrastructuur Technocratisch Weinig ruimte voor andere vakgebieden dan techniek
Fase II (1987 - 1998) In toenemende mate een autoritaire staat – loyaliteit aan het regime is wordt belangrijker De implementatie van integraal waterbeheer faalt	Fase II (1970 - 2002) De links georiënteerde overhead omarmt milieuwaarden Het maatschappelijk middenveld en belanghebbenden krijgen een grotere rol in besluitvorming
Fase III (1998 -) Reformatie leidt tot decentralisatie Toegenomen transparantie en verantwoording Wet nr. 7 over integraal waterbeheer wordt aangenomen Het belang van de competentie voor bestuur neemt toe	Fase III (2002 -) Privatisering om transparantie en effectiviteit van de overheid te verbeteren Organisatie wordt hiërarchischer

Ten tweede, op organisatie niveau moest de organisatie structuur in beide casus meer organisch worden in de overgang van Fase I naar Fase II, om een breder palet van vakgebieden en kennis te kunnen omarmen. Dit was een consequentie van het veranderende institutionele klimaat. In de Indonesische casus bleef de organisatiestructuur in essentie ongewijzigd, omdat er zoals gezegd op institutioneel niveau geen draagvlak was om naar een paradigma gestoeld op integraal waterbeheer te veranderen, met een grotere decentralisatie van besluitvormingsmacht. In Fase III kan de organisatie nog steeds gekenschetst worden door een hoog niveau van formalisatie en centralisatie van verantwoordelijkheden, wat uitgedrukt wordt in een sterke hiërarchie en strakke verdeling van taken en routines, ook al vraagt het waterbeheer in toenemende mate interdisciplinaire kennis en capaciteiten die buiten de organisatie gevonden moeten worden. Een meer organische structuur zou de uitwisseling van kennis met andere actoren vergemakkelijken, maar vraagt ook om erkenning dat die kennis niet alleen in de organisatie aanwezig is, en bovendien niet alleen aan de top. In de Nederlandse casus in Fase III, leidt de behoefte aan transparantie en (budget)verantwoording tot een meer mechanistische invulling van organisatiestructuur, met meer centrale controle over budgetten, werkplanning en personeelsbeheer (HRD). Dit heeft voor de verantwoording en transparantie naar publiek en politiek misschien goed gewerkt, maar het heeft ook een atmosfeer gecreëerd die minder bevorderlijk was voor het creëren en uitwisselen van kennis tussen staf, omdat zij bang waren initiatief te nemen.

Ten derde, op individueel niveau werd het duidelijk dat in de eerste fase en voor een groot deel ook in de tweede fase, beide casus een stevige technische oriëntatie laten zien, die in de Indonesische casus ook tot uiting kwam in de keuze voor technisch post-graduaats onderwijs door studenten. Tijdens Fasen II en III hebben in beide casus de ambtenaren een sterke voorkeur ontwikkeld voor het verkrijgen van administratieve en bestuurskundige vaardigheden, die er voor zorgen dat de balans met technische inhoudelijke kennis verloren is gegaan. Bestuurskundige competenties in de Nederlandse casus zijn in de tweede fase toegenomen, omdat de waterdeskundigen kennis en capaciteiten moesten ontwikkelen om met meerdere actoren om te gaan. In de Indonesische casus kon dit niet gebeuren omdat het politieke regime autoritairder werd, en loyaliteit beloonde ten koste van competentie. Beide casus onderschrijven het belang maar tegelijkertijd de lage waarde die gehecht wordt aan impliciete kennis. In beide casus gaat impliciete kennis verloren door vermarkting van taken en gebrek aan een goede planning van overdracht naar andere deskundigen, bij pensionering of overplaatsing. Tegelijkertijd geven alle respondenten aan dat de impliciete kennis bijzonder belangrijk is om het werk goed uit te voeren. De Indonesische casus laat zien dat de impliciete kennis het belangrijkste resultaat is van internationaal postgraduaats onderwijs. De studie toont aan dat impliciete kennis expliciete aandacht verdient in organisaties, in formele KCD mechanismen zoals onderwijs en training, door het voorzien in overdrachtsplanning, door het faciliteren van mentor-coach relaties, maar ook door informele mogelijkheden en een atmosfeer voor kennisuitwisseling te faciliteren. Ten slotte werd in deze studie bevestigd dat een conceptueel model voor KCD aandacht moet besteden aan het bestaan van de drie in elkaar genestelde niveaus: het individuele, organisatie en institutionele niveau. Bovendien moet het de relaties tussen de drie niveaus kunnen verklaren. In dit proefschrift heb ik een bestaand model veranderd en uitgebreid en heb ik dit aangevuld met een voorstelling van de relaties tussen de verschillende niveaus en de verschillende theorieën. Het aangepaste conceptuele KCD model kan samen met de nieuwe voorstelling toegepast worden als een structurerend en verklarend kader. Meer onderzoek wordt aanbevolen in andere sectoren en casus om de toepasbaarheid en volledigheid van het conceptuele KCD model en de nieuwe voorstelling te bevestigen. Onderzoek zou bijvoorbeeld kunnen focussen op het effect van persoonlijkheid en houding op individuele capaciteitsontwikkeling, leiderschap als persoonlijke competentie en als specifieke component van capaciteit in organisaties.

11 Annexes

ANNEX A: INTERVIEW QUESTIONS INDONESIAN CASE STUDY[57]

Standard questions for all interviewees:
Where, when and in which subject did you get your BSc
Where, when and in which subject did you get your MSc
What short courses outside the DGWR did you follow?

Tasks & competences

Describe your typical workday? What tasks do you have to do? How did you learn these things? Is there a difference in tasks between how it was before and after your post-graduate education?

What skills do you consider most important for your job? How did you acquire these?

Can you share some of your experience that is important for learning the business as a new PU employee?

(for people who lead a team:) What people do you prefer to hire in your team? What kind of factual knowledge and skills and attitude do they have?

(If no management function:) What combination of knowledge and skills do you think are necessary to fulfil tasks in a team?

Do you know how many people have a technical background in a management function at PU?

Who are your heroes/examples in your organisation, in the technical field, management, governance? Why is it, what characteristics do these heroes/examples have? How are they perceived in the organisation?

What was the most challenging task you have had in your career? Were you prepared for it?
What type of tasks do you like best?
What are the most challenging issues you face in daily working life?

KCD mechanisms

How do you keep up to date, how do you acquire new knowledge about your profession? What do you read? How often do you read work related material?
Do you often see things on TV that are related to your work?
Do you have discussions with colleagues about work? In which context do you have these discussions? (e.g. in break time, or during project meetings)
Do you know of professional networks in Indonesia and are you a member of such a network?

---

[57] Translated from Indonesian

If you use email for work related matters, how fast do you expect an answer?

What is your drive/motivation/inspiration for your work?

Post-graduate education
What did you learn at your post-graduate education, other than technical knowledge/skills?

What did you expect to learn at local post-graduate education, and international post-graduate education (if applicable), how useful was it for your work, in general?

What would have happened if you never did the post-graduate education outside Indonesia?

How appropriate was the level of technical knowledge you gained in your post-graduate education?

Did you acquire management skills during the post-graduate education? Communication, negotiation, planning skills, proposal writing, financial management?

Did you acquire governance skills during post-graduate education? The ability to engage with different stakeholders, decision making skills etc?

Did you learn (how) to reflect on your own performance?

What was most pleasant and what was most frustrating in your post-graduate education?

To what extent was your education connected to the content of your job?

Did your education change your perspectives and behaviour, and how?
Can you see a similar trend with friends/colleagues?

Did your educators try to develop a certain attitude in the students?

When you took your education, was it focused on classroom learning, peer learning, learning by doing etc? (Name different learning vehicles as an example)

Do you see differences in how people work, based on where they studied?

About the organisation:

What incentives are present in the organisation to acquire knowledge and skills?

How were you selected to study abroad? Was it your own initiative, were you appointed by your superior, or was it linked to a performance evaluation system?

Do you see changes in how people acquire their knowledge over the years?

How would you describe the working culture in your organisation? Can you give an example that describes the ways of the organisation?

What was the follow-up when you came back from your studies abroad?
What would you have liked to happen, as follow up from your education abroad?
How would you help your colleague/friend when he/she comes back from education abroad?
Should there be a role for the educational institute in the follow-up?

Do you have, or have you ever had a mentor or coach for your work, or are you a mentor yourself?

Is education or training available, facilitated by your organisation to update your knowledge? To what extent is it connected to the contents of your job?
Does your organisation have financial resources to support learning by staff?
Does your organisation have procedures to support learning by staff? What are the instruments for that purpose?
Can you make use of these procedures?
Does your organisation have procedures to assess your performance?

What do you identify as most important outside and inside influences on the knowledge and capacity in your organisation? Think of technology development, attention for social issues, larger changes in society? Do you see the outside changes reflected in your organisation?

Can you describe the type of person that is successful in your organisation? What are the assets that you need?

How did the knowledge and capacity needs of the organisation change over time? How was it in the '80s, in the '90s and after 2000?

How are the changes of law 7/2004 reflected in the training of PU employees? What new skills and knowledge does it require? How do employees acquire these skills?

How do recruitment policies for the government in general, influence the recruitment in PU? How do you think about the zero-hiring policy of the 1990s?

What do you think about outsourcing (hiring of external consultants to do the job) of specialist activities to the private sector, if you consider knowledge and capacities in this organisation?

ANNEX B: QUESTIONNAIRE INDONESIAN CASE STUDY[58]

PERSONAL INFORMATION
Personal data (anonymous but some details are needed for statistical analysis)
Gender   M ☐   F ☐
Year of birth: ....

EDUCATION
Highest level of education (please encircle the correct option):

BSc / MSc / PhD / Other

Field of study:

Civil Engineering / Environmental Engineering, Ecology or Biology / Hydrology/ Integrated Water Resources Management Economics / Law / Public administration/ Other :
....
....
....

Year degree was obtained (please fill in)
....

From which institution did you obtain your highest degree (please encircle the correct option, and in case not listed, write it down)?

Local:
Gajah Madah University
University of Indonesia
University of Sri Wijaya
Institute of Technology Bandung
Institute of Technology Surabaya
Other:
Delft University of Technology, NL
University of Grenoble
University of Lyon

international:
Colorado State University, USA
University of Manitoba, CA
University of Roorkee/Indian Institute of
Technology, Roorkee, IN
Asian Institute of Technology, Bangkok, TH
(Unesco-)IHE, NL

Other:...

---

[58] Translated from Indonesian

CAREER INFORMATION

Please enter the all job posts you have had in chronological order, starting with your current job post and ending with your first job. In case you have had more than one job at the same time, only enter that job that provided you with the most income. After a major change/shift within the same job or organisation (i.e. you got promoted, transferred to another country), please treat this as new a job/job position on a new table row.

	Job Level (Rank nr. and letter) & name of position	DG & Directorate	Sub-directorate	Main activity (Choose from:) 1. Information & knowledge dissemination 2. Project implementation & supervision 3. Management 4. Research 5. Policy formulation 6. Project design 7. Procurement 8. Contracting 9. Planning	Start year	End year
Job/job position 1						
Job/job position 2						
Job/job position 3						

194

Job/job position 4					
Job/job position 5					
Job/job position 6					
Etc.					

By how much did your total income (including bonuses) increase: (to avoid confusion: 100% represents your salary at the time of obtaining your highest degree, so 150% is 1.5 times your salary at the time you obtained your degree, 75% represents a salary reduction of 25%. If you didn't work before obtaining your highest degree, fill in the amount after 1 year as 100% and compare it with your salary after 5 years.)

1 year after obtaining your highest degree?   ☐ 50%   ☐ 75%   ☐ 100%   ☐ 150%   ☐ 200%
Other: ....
5 years after obtaining your highest degree?   ☐ 50%   ☐ 75%   ☐ 100%   ☐ 150%   ☐ 200%
Other:..................

To what extent did you obtain the following competences during the education for your highest degree and to what extent are they required in your current job. If you are currently not employed please answer only column A.

During your highest education						used in your current job				
Not At all				Extensively		Not at all				Extensively

Technical competence:

1	2	3	4	5		1	2	3	4	5
☐	☐	☐	☐	☐	Modelling skills	☐	☐	☐	☐	☐
☐	☐	☐	☐	☐	Computational skills	☐	☐	☐	☐	☐
☐	☐	☐	☐	☐	Design skills	☐	☐	☐	☐	☐
☐	☐	☐	☐	☐	Subject-specific theoretical knowledge	☐	☐	☐	☐	☐
☐	☐	☐	☐	☐	Field-specific knowledge of methods	☐	☐	☐	☐	☐
☐	☐	☐	☐	☐	Understanding of broader technical context	☐	☐	☐	☐	☐
☐	☐	☐	☐	☐	Multidisciplinary thinking/knowledge	☐	☐	☐	☐	☐
☐	☐	☐	☐	☐	Analytical competence	☐	☐	☐	☐	☐
☐	☐	☐	☐	☐	Problem solving skills (technical)	☐	☐	☐	☐	☐
☐	☐	☐	☐	☐	Practical field based knowledge	☐	☐	☐	☐	☐

Management competence:

☐	☐	☐	☐	☐	Project management skills	☐	☐	☐	☐	☐
☐	☐	☐	☐	☐	Networking skills	☐	☐	☐	☐	☐
☐	☐	☐	☐	☐	Negotiation skills	☐	☐	☐	☐	☐
☐	☐	☐	☐	☐	Procurement	☐	☐	☐	☐	☐
☐	☐	☐	☐	☐	Contracting	☐	☐	☐	☐	☐
☐	☐	☐	☐	☐	Time management	☐	☐	☐	☐	☐
☐	☐	☐	☐	☐	Team work	☐	☐	☐	☐	☐
☐	☐	☐	☐	☐	Problem solving skills (managerial)	☐	☐	☐	☐	☐
☐	☐	☐	☐	☐	Oral communication skills	☐	☐	☐	☐	☐
☐	☐	☐	☐	☐	Written communication skills	☐	☐	☐	☐	☐
☐	☐	☐	☐	☐	Leadership skills	☐	☐	☐	☐	☐
☐	☐	☐	☐	☐	Personnel management	☐	☐	☐	☐	☐
☐	☐	☐	☐	☐	Organisational management	☐	☐	☐	☐	☐

Governance competence:

☐	☐	☐	☐	☐	Ability to apply inclusiveness	☐	☐	☐	☐	☐
☐	☐	☐	☐	☐	Policy formulation skills	☐	☐	☐	☐	☐
☐	☐	☐	☐	☐	Understanding of procedures and Institutional structures	☐	☐	☐	☐	☐
☐	☐	☐	☐	☐	Understanding of political consensus building	☐	☐	☐	☐	☐
☐	☐	☐	☐	☐	Achieving ethical objectives: non-corruption, transparency etc.	☐	☐	☐	☐	☐

Learning competence:

☐	☐	☐	☐	☐	Reflective thinking, assessing one's own work	☐	☐	☐	☐	☐
☐	☐	☐	☐	☐	Creativity	☐	☐	☐	☐	☐
☐	☐	☐	☐	☐	Critical thinking	☐	☐	☐	☐	☐
☐	☐	☐	☐	☐	Intercultural understanding	☐	☐	☐	☐	☐
☐	☐	☐	☐	☐	Broader view of the world	☐	☐	☐	☐	☐

☐☐☐☐☐ Curiosity   ☐☐☐☐☐

What type of knowledge and skills is valued most by your direct superior? ------------------------
----

TRAINING

Have you undertaken **non-degree** training? If yes, please specify the approximate total number of weeks or months of training taken.

☐   No
☐   Yes, how many weeks: ....

Please enter the training events that were most useful for your daily work.

	Topic	Training given by	Country	Start	End
	(For example: IWRM, financial management)	(for example: pusdiklat, HATHI, etc.)		(month /year)	(month /year)
major training event 1					
major training event 2					
major training event 3					
major training event 4					
major training event 5					
remarks					

OTHER KNOWLEDGE ACQUISITION MECHANISMS

In our jobs, we use different sources and ways to acquire knowledge. Can you indicate what sources you make use of to keep your knowledge up to date, and to what extent.

	Not at all				Extensively
	1	2	3	4	5
Training	☐	☐	☐	☐	☐
Seminars/workshops	☐	☐	☐	☐	☐
Formal work related meetings	☐	☐	☐	☐	☐
Informal meetings with colleagues	☐	☐	☐	☐	☐
Having a senior mentor/coach	☐	☐	☐	☐	☐
Interaction with foreign consultants	☐	☐	☐	☐	☐
Internships/job shadowing	☐	☐	☐	☐	☐
Learning by doing	☐	☐	☐	☐	☐
Networks (professional and personal)	☐	☐	☐	☐	☐
Network organisations (such as professional associations)	☐	☐	☐	☐	☐
Online networking (facebook, LinkedIn)	☐	☐	☐	☐	☐
Databases of your organisation	☐	☐	☐	☐	☐
Mailing lists inside organisation	☐	☐	☐	☐	☐
Mailing lists outside organisation	☐	☐	☐	☐	☐
Academic literature	☐	☐	☐	☐	☐
Relevant newsletters / magazines inside organisation	☐	☐	☐	☐	☐
Relevant newsletters / magazines outside organisation	☐	☐	☐	☐	☐
Media - TV/Radio/Newspapers	☐	☐	☐	☐	☐
Open source sharing (wikipedia, flickr, youtube)	☐	☐	☐	☐	☐

Comments:

....

....

To which extent do you consider your current professional activities related to the development of:

	Not at all				Extensively
	1	2	3	4	5
Your organisation?	☐	☐	☐	☐	☐
Your country?	☐	☐	☐	☐	☐
Other countries comparable to Indonesia?	☐	☐	☐	☐	☐

If you studied abroad during your career in Public Works, to what extent was your working environment conducive for using the new knowledge and skills upon return?

If it was conducive, were you encouraged to share your knowledge through:

	Not at all				Exten-sively
	1	2	3	4	5
Presentations for other staff	☐	☐	☐	☐	☐
Through becoming a mentor of junior staff	☐	☐	☐	☐	☐
Specifically asked to give your input in organisational matters	☐	☐	☐	☐	☐
Tasks allocated that require your specific new knowledge and skills	☐	☐	☐	☐	☐
Further training possibilities as follow up and refreshment of your education abroad	☐	☐	☐	☐	☐

REMARKS:

....

....

....

....

....

....

....

....

....

....

....

....

....

ANNEX C: INTERVIEW QUESTIONS DUTCH CASE STUDY[59]

What is your vision on knowledge and capacity development in the Dutch public water sector?

What do you see as distinguishing moments in the development of Dutch water management?
By what factors are these moments determined?

How would you describe the interaction between various players in the sector? Between public and private, between public and knowledge institutes and universities?

How does the sector acquire knowledge?
Would you say that happens through trial and error, through a well designed strategy?
Is it a combination of all kinds of factors?
What kind of knowledge is important, e.g. experience vs. explicit knowledge, attitudes and skills?

Technical knowledge is very important, but knowledge of management and governance as well, to broaden the field. How would you rate the competence for water governance (explain)?

What determines development in the public water sector? How does the functioning of the water sector improve? How do you see the role of knowledge in that development process?

How do you think about outsourcing of work from the government to the bureaus?
How are outsourced assignments monitored and supervised?

How did academics play a role in KCD in the Dutch water sector? What should be their role?
Is what students learn at the TU Delft well connected to what is needed in the sector?
How was this approximately 30 years ago? How did the curricula at the TU Delft change over time?

---

[59] Translated from Dutch

ANNEX D: QUESTIONNAIRE DUTCH CASE STUDY[60]

1.   Are you male or female?

2.   What is your year of birth?

3.   For which company or organisation are you currently working?

4.   Can you indicate which description fits your organisation best in the years 1990, 2000 en 2010? We would also like to ask you to make an estimation for the year 2015. In what type of organisation do you think you will work in 2015? Please leave the field blank if you were not working in certain years.

	1990	2000	2010	2015
Local government incl. water boards				
Provincial government				
National government				
Knowledge institute incl. academia				
International organisation				
Non-governmental organisation				
Drinking Water utility (private sector)				
Construction (private sector)				
Consultancy (private sector)				

[60] Translated from Dutch and originally formatted in SurveyMonkey

5.   In which sub-sector are you or were you working?

	1990	2000	2010	2015
Drinking water supply				
Water and wastewater treatment				
Agricultural water management				
Construction of water related infrastructure				
Water quality management				
Water quantity management				
Integrated water management				
Other				

6.   What is your current position in your organisation and what position did you hold in earlier years. Please give an indication for 2015:

	1990	2000	2010	2015
Leadership				
Policy formulation				
Implementation				
Support				
Research				
Other:				

7.   What is your highest obtained degree? Choose from: BSc/MSc/PhD

8.   Please answer the following questions about your highest obtained degree:
      What is the name of the study programme?
      What is the name and location of the institute?
      In which year did you obtain your highest obtained degree?

9.   How did your education shape your view of the world?

10.   How did your education connect to your first job? Please indicate per component to what extent it was sufficient for the demands in your first job.
Give a score from 1 to 5, where 1 = not at all, 3 = to some extent, and 5 = to a large extent.

	Score
Technical substantive knowledge	
Understanding of the wider context	
Skills	
Attitude	
Believes and values	

11.   Can you indicate to what extent the following components of knowledge were important to you in your career, in 1990, 2000 and 2010? Please indicate for 2015. Give a score from 1 to 5, where 1 = not at all, 3 = to some extent, and 5 = to a large extent.

	1990	2000	2010	2015
Technical substantive knowledge				
Understanding of the wider context				
Skills				
Attitude				
Believes and values				

12. How many hours per week are you in contact with other water professionals, including your own colleagues? (e.g. through meetings, e-mail, phone, networking a.o.) Can you indicate this for 1990, 2000, 2010 and can you give a prediction for 2015 by checking the appropriate box?

	1990	2000	2010	2015
< 1 hour				
1 – 5 hours				
5 – 10 hours				
10 - 15 hours				
15 – 20 hours				
25 – 30 hours				
35 – 40 hours				
40 - >				

13. Please indicate for 1990, 2000, 2010 and estimate for 2015 which means you use to acquire knowledge that is necessary for your job? Please give a score from 1 to 5, where 1 = not at all, 3 = to some extent and 5 = extensively

	1990	2000	2010	2015
Training/course				
Seminar/workshops				
Project related meetings				
Through a mentor/coach				
Job shadowing				
Learning by doing				
Through your personal/professional network				
Through network organisations				
Online networking				
Databases in your organisation				
Reading academic literature				
Relevant newsletters/ magazines				
Media: TV/Radio/Newspapers				
Online open source material				

Comments:

14. Stakeholders are individuals or groups that are directly or indirectly involved in a specific topic, problem or activity. How does your organisation involve those stakeholders to make use of the knowledge of those stakeholders, in 1990, 2000, 2010 and what do you predict for 2015?

	1990	2000	2010	2015
We inform stakeholders				
We consult stakeholders				
We co-produce with stakeholders				
We make decisions with stakeholders				

Comments:

15.   Can you indicate for 1990, 2000, 2010 to what extent knowledge is lost in your organisation because of the reasons mentioned in the Table? Please give an estimate for 2015.

	1990	2000	2010	2015
Due to lack of communication between stakeholders				
Because of outsourcing from public to private sector				
Because of retirement of experts				
Because of a lack of succession planning				
Because process management has become more prominent in your organisation				

Comments:

ANNEX E: THEMATIC AND OPEN CODING USED IN ATLAS.TI

Thematic codes	Open codes
Accountability frame	Appropriate training and education while on the job
Administrative Procedures	
Apprenticeship	Availability of information
Attitude	Construction paradigm
Budget frame	Corruption
Change Management	Critical reflection on organisational performance
Civil society	
Competence for continuous learning and innovation	Focus on technical competence vs. interest for multidisciplinary approach
Coordination and collaboration	Formal
Dialogue with stakeholders	Incentives
Education	Informal
Factual knowledge	Instruments to keep up to date
Fiscal frame	Local vs. international education
Governance competence	Preference for certain schools
HRD	Relationship boss – employee
HRM	Selection procedures for education and training abroad
Individual level	
Institutional level	Social learning
Learning by doing	Social networks
Management competence	Status
Networking	Teaching style
Organisational level	Theory vs. practical experience
Organisational strategy	Working culture
Peer learning	
Performance	
Policies	
Political influence	
Press	
Public administration	
Regulatory frame	
Skills	
System to learn lessons	
Technical assistance	
Technical advice on organisational structure, management and incentives	
Technical competence	
Training	
Transitions	
Triggers	
Understanding	

4: PD46: 1972: in the subdirectorate for irrigation, students were very theoretical, they were unprepared for practice. The only practice they had was their project for their bachelor and master degree, of three months. So upon arrival in the ministry they had to do a real design, guided by staff from the ministry, bring it to the field and construct it. Then they knew better how things worked in practice and had more confidence. In such a period it is also possible to see the character of the person. There they selected where the new persons would be posted. They left them there for a year and then rotate. This approach was adopted by swamps and river dir and lasted until 1990. This was a very effective approach.

1979: The workload of the people became higher and higher, they started to delegate to the province. They set up a design unit at that level. Young engineers in that group worked together with dispatched people from the central level, to learn. At that time there were no lecturers in construction or construction management in ITB. In the design unit there was assistance from international donors, consultants.

PD57: We had a motto from the minister in 1965: work hard, act fast, and be precise, take the right decisions. That was the culture, and we all tried to do that, because it was monitored by the minister. I don't know if it is still successful, maybe partly. The motto is still valid. Some people work that way, others don't. I see people working till late, but others are reading the newspaper. There are now a lot of meetings, more than in my time. Civil servants have to prepare questions for the parliament; they're a very strong force now. In my time it was not like that. Now they can ask the DG to come and ask him questions.

5: PD32: By the end of the 1980s, beginning of 1990s, the knowledge and expertise decreased compared to earlier years.

PD48: They have to start again like in the earlier days, when young officials worked in the field with older engineers. This provided a transfer of knowledge.

7: PD37: Political connections are important in Indonesia, if you want to get somewhere.

PD51: Should I mention names? Usually in PU you have to choose a political colour, a leader. So if I choose a name, and I am afraid that you relate it to a party? Actually I didn't join anything, but probably in my time of service, later, I have to choose. I hope I don't have to. In Public Works, under this president, it is impossible not to choose.

PD53: Most of our alumni have a good position now. But formally there is no follow-up, but we can see that they are generally doing good. Especially at the central level. For local people it is rather difficult. It is different there. The political affiliation is much more important. There are many jobs but a lack of persons, and often when we propose a person with the right qualifications, this person will be used for something else.

PD57: Personal relations between ministers and DGS are very important in Indonesia. The ministers and DGs they need to have strong relations. Now there is too much party politics. In my time i had friendly relationships, so it was easier, the political situation was different.

8: PD27: PU is a very old organisation, based on seniority, rank, position, and sometimes you're not allowed to talk. I am not a person who wants to be held like that. My two bosses always say, people know you if you're speaking, but not all senior staff like that. I also heard people saying,"if he would be my staff I would fire him".

PD32: the planning of work is done hierarchical; there is no discussion about the planning. It is done based on assignments from higher levels, or to solve problems.

PD35: Things are changing in minPW: the newer people are better educated than before; procedures to get a job at minPW are easier now. Communication with senior staff is still difficult. Middle management is no problem.

PD36: He was in a different world, he says. The director and driver could sit at one table. Here in Indonesia we still live in feudal times. There you could also show weakness. When he came back he had clashes with his boss.

PD38: Young people are not permitted to grow, because of the one man show. If you're young and want to grow, this will depend on your boss.

PD42: In Indonesia people look to the seniors, to the parents. There is lots of respect for older people. But this is sometimes a false perspective of respect. ABS= asal bapak senang: not focusing on right or wrong, but to make superior happy.

PD50: We are always afraid of the people above us. There is manipulation. That is important to improve.

PD58: You do not need to be clever in the sense of clever in school; you need a different type of cleverness. How to work in such a system. PU is quite a unique place; there is a system, a way of working. You cannot talk frankly and you cannot push. So in the beginning I asked everyone, ok how do you do this, or how do you do that. I asked mostly colleagues at the same level, or my staff, but not my boss.

PD62: Employees are still determined by their superior, so they lack initiative and creativity. They lean to what their boss says, and they follow his orders. This goes for most of the employees. We have other constraints, that if you take too much initiative, if you are too creative, you are facing problems with your boss, because he will think that it's not appropriate for you to be his staff, and maybe you will be ignored by your boss, because he may think you want to show him you are better, and have more skills and knowledge than he or she has. So better be careful with this.

9: PD51: There is not much change over the years. I have big hopes for the future generation. I push the younger staff so much; I give them motivation, to do more, and to change the hiring system.

PD62: I have a lot of discussions with older people, the older generation. I am learning how to address my critical thinking to them politely, in a way they can accept it. I am still in the process.

10: PD52: The director gave me the job, so I try it. In private companies it's not like this, but in the government, we just endure it. No problem, we try to go along with it. There are many here that studied something, but do something totally different in their work. No match.

11: PD27: Every day in IHE we talk about IWRM. We have the concept here, but it is not common knowledge. We have a program related to IWRM but staff like me never worked with it.

I think it is one of the biggest advantages to learn in IHE, because I got broader views. In IHE it is not totally technical, especially in water resources management. I got sociology, it was so confusing, I hated it :) at first, but later I tried to enjoy it. You know, it is difficult to start like that, compared to just reading books. We learned about international conflicts, transboundary, about stereotypes, Hofstede. Maybe if I got similar lectures from Indonesian lecturers maybe I just think ok, perhaps that is the way we see them. But if you hear other people say that about you, and other people from other countries, than you think, maybe this is true.

PD40: When coming back from Delft he was doing things differently. In delft he learned to plan ahead. Small changes that he adopted. He liked the level of regulations and organization in the Netherlands. It can only be adopted unto a certain level.

PD51: Open mind and critical thinking. In our culture, some people get offended if you criticize. I am supposed to be critical, but I cannot use it here. You change inside, but I cannot apply it here. My behaviour is very pragmatic, but I am changed inside, I don't want to implement it, I cannot implement it, but I know it should be done. You know, with critical thinking you find a lot of solutions, alternatives.

PD64: ...it changes your perspective, broadens your horizon, especially when I got my masters. I could see the world; I saw people from all over the world and learned experiences from all over the world. It was transforming me in someone with more knowledge, and perspective of life in general and also in how to do engineering work. Particularly, basically my interest was to become an architect. And I couldn't because if I wanted to be an architect I had to go out of my town and as my sister is also going to university and the expenses would be too high if I go out of town. So I chose to join civil engineering.

I do feel that my higher education changed me.

I found that people from other countries show the same type of people in Indonesia. You have the stubborn one, the cooperative one, the silent one; I learned that it is universal.

And other things, like that it is dark when you wake up, and you go home and it is already dark. The seasons make you more disciplined.

12: PD27: I was single and far from my family and my mother wanted me to be settled. What kind of job could I have that allows me to get settled. At that time I applied for local government but if I could achieve something more, then why not.

PD35: He likes the security that minPW provides.

PD37: They are safety players. The ministry needs to become more opportunistic, in a positive way, and entrepreneurial. They always stick to rules and regulations.

13: PD48: The mindset of the minPW is very different than at the ministry of agriculture. For the minPW interaction with local actors is not interesting. It is not because they are arrogant, but just because they are not interested.

PD49: ...There is a lack of communication, collaboration between institutions. Between institutions, between ministries, between DG's.

15: PD50: For example me, after i entered this office, I was sent to a short course, but the subject of the short course was different than my daily tasks. School and courses are very different from what you do in your daily tasks.

PD52: Also goes for the training in Public Works, some training that is organised matches, some doesn't. If it doesn't match I just go to listen, I just come along.

PD54: Some of them (alumni) have learned many things abroad, but cannot apply it in practice, they cannot speak about it.

16: PD49:...there are too many civil servants working on administration. All studies or consultancy services are done by third-parties, outsiders, outsourced. For the young generation, and the existing personnel, they only get the report/results of what the consultant has done.

PD58: Why we don't have time? Because we have to manage administrative things. We don't talk about technical things, we forget about that, because we're only busy with the management.

17: PD27: Something like coordination or collaboration with other parties is less important.

PD35: Communication with senior staff is still difficult.

PD48: The mindset of the minPW is very different than at the ministry of agriculture. For the minPW interaction with local actors is not interesting. It is not because they are arrogant, but just because they are not interested.

PD49: ...There is a lack of communication, collaboration between institutions. Between institutions, between ministries, between DG's.

PD50: We need more output from the organization, and coordination, collaboration.

PD56: Difficult, still. There's still a lot of sectoral egoism. it is now better than in the 1960s or 70s, but it is still there.

18: PD38: Young people are not permitted to grow, because of the one man show.
If you're young and want to grow, this will depend on your boss.

PD39: After his studies abroad he tried to diminish the hierarchy, because discussions can be easier, and people can have an open mind. In a research department the hierarchy is anyway less.

PD56: My staff is mostly updated by me, if I find something interesting, I share it with my staff. At the moment it works like this.

19: PD39: Most people who come in have basic knowledge in their field, at the level of BSc, but they are not ready to go to the field. They have to be attached to a senior.

PD46: The design unit produced capable engineers. There was day to day guidance.

PD48: They have to start again like in the earlier days, when young officials worked in the field with older engineers. This provided a transfer of knowledge.

PD49: He got the task from the DG and the consultant, to guide the juniors as a mentor, to improve their knowledge.

20: PD32: Every task in the projects was outsourced to local consultants. Expert knowledge became less relevant and people who were good administrators went up in the organization. The idea was: "If you need something, hire a consultant."In this way you lose a lot of knowledge and history.

PD36: Capabilities of Public Works are decreasing because of outsourcing. There is nothing for engineers, and they have no field experience.

PD47: The ministry now does a lot of outsourcing, but the quality of that work is lower.

PD48: Outsourcing of construction works is ok, but Public Works has to supervise it. It is not able to do this.

PD49: All studies or consultancy services are done by third-parties, outsiders, outsourced. The young generation only receives the report/results of what the consultant has done.

PD53: Technically the ministry is not growing, because of outsourcing. The outsourcing started in 1990. Before that work was done in force account.

22: PD30: Generally, people in Public Works are more interested in technical topics.

PD31: Communication was clear, because everything was about engineering.

PD32: People with a good overview are needed, but they do need an engineering background.

PD33: Also Indonesian institutions signalled problems due to the fact that the ministry of public works focused solely on infrastructure development.

PD41: People need technical knowledge and need to have a technical background, together with soft skills.

PD64: Managers should always have a technical background.

23: PD27: I take this as a chance, a new experience. I am sure I will have an opportunity again if I want to see huge constructions.

PD31: It is important for graduates to learn all the time.

PD38: People need a motivation to learn. This should also come from the organization, the organization should encourage learning.

PD46: He likes to try new things, and when they work he likes to include his friends in these developments.

PD56: I get it from books, workshops, seminars. i read when i need it. or use the internet. I like to learn from any source.

PD58: I was afraid to talk in the beginning, so I thought I should start to develop myself. Now I like to talk to seniors, to make my opinion clear. My direct boss is open-minded.

24: PD27: For me it was quite satisfactory, my foreign education, something totally new for me. If I want to learn how to build a dam, I can just buy a book and read it.

PD27: ...We had the opportunity to go to Delft and we chose water management for a reason. I thought that technical things are for technical staff, and we wanted something more, we wanted some comprehension about water issues, that's why we chose water resources management. We believed that we could get technical content at Public Works, we have training for that.

PD63: I have my background in civil engineering, but we need to know the regulations, laws, public administration and management. Our tasks are quite varied.

PD51: ...in government you govern, or manage. So the type of leader that I like to see is someone who can manage, and has not only the technical skills. Technical skills are good for your foundation, because this is a technical department. For the rest it is all management, but you need the core. Once you grow in this dept, you need management, good adaptability, good balancing skills, and a good approach to people.

25: PD27: Every day in IHE we talked about IWRM. We have the concept here but, it's not common knowledge. We have a program related to IWRM but staff like me never worked with it.
I think it is one of the biggest advantages to learn in IHE, that I got broader views. In IHE it is not totally technical, especially in water resources management. I got sociology, it was so confusing, I hated it :) at first, but later I tried to enjoy it. You know, it is difficult to start like that, compared to just reading books. We learned about international conflicts, transboundary water management, about stereotypes, Hofstede. Maybe if I got similar lectures from Indonesian lecturers maybe I just think ok, perhaps that is the way we see them. But if you hear other people say that about you, and other people from other countries, than you think, maybe this is true.
PD51: Open mind and critical thinking. In our culture, some people get offended if you criticize. I am supposed to be critical, but I cannot use it here. You change inside, but cannot apply it here. My behaviour is very pragmatic, but I am changed inside, I don't want to implement it, I cannot implement it, but I know it should be done. You know, with critical thinking you find a lot of solutions, alternatives.
PD54: Someone who studies outside the country will be in touch with other ideas and can mix that with the experience he already has. The contribution is technical, with new expertise, but also to be open to new ideas.

26:PD32 and 33: During the Suharto regime you had to be loyal to make a career. You needed friends in the right places.

27: PD33: Also Indonesian institutions signalled problems in the fact that the ministry of public works focused solely on infrastructure development. Up to today this is still the case; the ministry is hardly looking at other proficiencies than engineering.

28: PD27/PD62: ...We had the opportunity to go to Delft and we chose water management for a reason. I thought that technical things were for technical staff, and we wanted something more, we wanted some comprehension about water issues, that's why we chose water resources management. We believed that we could get technical content at Public Works, we have training for that.

30: PD50: What you learn in university is very different from daily work, it is very theoretical.
PD58: Practice is very different from what I had in school, much more theoretical.

31:PD27: There are 3 major things I got from IHE. First thing is a more broadened field: international circumstances with different people and opportunity to see the south of

Spain, the world water problems in arid areas, conflicts with the Evoradam. That was amazing.

PD38: He was lucky to go and study in Delft, it broadened his view. He learned how to communicate, how to analyse problems. Theory was connected to practice.

32: PD27: We got those, communication, negotiation, planning, proposal writing. Not too many financial skills.

PD56: In my Master of Management education for my bachelor degree, I learned communication skills, negotiation, etc.

33: PD27: After 2005 we had a lot of eyes, auditors and such, and people were saying, don't let this look wrong in the eyes of the auditors. So there was less discussion about technical stuff, more about how the work should look good in the eyes of the auditors. So they left the technical stuff to the working units. In my opinion: why don't we talk about technical problems all the time?

PD50: You have to look at the real problem. Not only at symptoms, or problems resulting from the real problem. Look at what is behind it. Evaluations must give recommendations to repair the fundamental problem, not only symptoms. Leaders should talk about this with their colleagues in the field.

34: PD38: He learned how to communicate. He remembers the easy discussions with others from other disciplines.

PD41: The period there made him change. Not only technical knowledge, connected to other disciplines, but a broadening of his view stimulated by a different culture, by the way people think and live.

35: PD27: I take this as a chance, a new experience. I am sure I will have an opportunity again if I want to see huge constructions.

PD38: People need a motivation to learn. This should also come from the organization, the organization should encourage learning.

36: cited in text

37: cited in text

38: PD42: He believes that by studying abroad the mindset of people doesn't change a lot. Their education there should be together with people from NGO's, university staff, not only people from the Public Works. They are still "in the box" after their studies.

39: PD70: People were shocked to notice that dead fish were floating in the Rhine. Not only local, but from Mainz in Germany to the Netherlands. And at some point in 1971 it was not possible to take water from the rivers to process drinking water, as it was too polluted. Protest groups were established to protest against these developments.

PD71: Protest groups played a very important role in this shift, and they in turn were there because a general awareness was growing in society that we needed to work on environmental issues.

41: PD71: Those protest groups were the conscience of society.

42:PD70: They taught us to think in alternative solutions.
PD71: Rand Corporation transferred a lot of knowledge. Part of it was Human Resource Development, how to motivate and coach people.
PD73: De eerste keer dat er een risico analyse werd gemaakt, een onzekerheidsschets werd gemaakt. Kansen berekenen. Kwam van wetenschap over vliegen naar de maan, daar deden ze dat.
PD73: It was the first time a risk analysis was done. Calculating chances, getting information about risks and insecurities. This originated in the knowledge about flying to the moon, that's where the experience came from.
PD74: That was a very useful study, generating a lot of knowledge.

43: PD71: A specific approach per person. Weak points were developed. It was a kind of coaching.

44: PD70: After the 1960s it became more complex, the environmental aspect became important, the international dimension, fish, biodiversity and drinking water security. And on the political side the stakeholders that wanted to be informed.
PD71 protest groups played a very important role in this shift, and they in turn were there because a general awareness was growing in society that we needed to work on environmental issues. Society also wanted to be informed and involved.

45:PD74: The spatial planners did not agree that the water managers want to have a say in the use of space. There is a tension, you have to draw boundaries somewhere in reality, to manage it, and then some kind of wall is erected. To tear it down you need an external trigger.

46: PD74: We could cooperate with other ministries as we deemed necessary, there was no need to ask permission.

47: PD74: Everything had to go past the management team of the Rijkswaterstaat. This is a ridiculous situation, and what you will get as well is fear. The management team was harsh to the employees, so people will not stick their neck out.

48: PD47: If employees stay too long in one function, the assumption may be that they are not capable.

49: PD70: This is how I want to show that the current tendency to outsource all work to the private sector yields a lot of fragmented knowledge. The government is actually erasing its memory by doing so.

PD71: Yes, we have lost our memory now.

PD71: The Rijkswaterstaat has been manoeuvred in an impossible position. You (the Rijkswaterstaat) have to give an assignment, for which you require knowledge. You do not get the people who have the knowledge to give these assignments. So the assignment is often imperfect. The implementation and preparation has to be handed over to the private sector, but you are in the end responsible for the quality. The people who are knowledgeable are either fired or retired.

PD72: In the past the Directorate Generals went through the whole system, and knew the local conditions. They had feeling for the substance. They would also guard a minister for incorrect statements. This is now completely gone.

50: PD71: The Rijkswaterstaat has been manoeuvred in an impossible position. You (the Rijkswaterstaat) have to give an assignment, for which you require knowledge. You do not get the people who have the knowledge to give these assignments. So the assignment is often imperfect. The implementation and preparation has to be handed over to the private sector, but you are in the end responsible for the quality. The people who are knowledgeable are either fired or retired.

51: PD74: cited in text

52: PD71: The Rijkswaterstaat has been manoeuvred in an impossible position. You (the Rijkswaterstaat) have to give an assignment, for which you require knowledge. You do not get the people who have the knowledge to give these assignments. So the assignment is often imperfect. The implementation and preparation has to be handed over to the private sector, but you are in the end responsible for the quality. The people who are knowledgeable are either fired or retired.

PD73: the Rijkswaterstaat used to have a lot of expertise in the organisation. Such as contractors, they were state companies. Now many tasks are outsourced, and the knowledge to supervise the work is gone as well.

53: PD74: cited in text

54: PD74: You see that many younger people do a traineeship, they see the whole organisation, come and talk to people, follow courses. This is a bit like a coaching trajectory.

55: see footnote 48 to 52.

56: see footnote 48 to 51

12 Acknowledgements

When I graduated in 2001 from Wageningen University in Irrigation and Water Engineering, I wanted to see the world. In my first job as a consultant for Arcadis Euroconsult, this wish became reality and I worked and lived in Yemen, Indonesia and Egypt for three years. At that time I had only vague, undeveloped thoughts about pursuing a PhD. The work acquainted me with different players in the public water sector and a variety of water management problems. I enjoyed this experience tremendously. After this period, I wanted to experience a different approach to development projects and I started working for ICCO (Interchurch Organisation for Development Cooperation) in 2006. In this job I worked with many local partner organisations throughout Indonesia and in Timor Leste, ranging from rehabilitation work in the regions hit by the tsunami in 2004, to drinking water projects in West-Papua. The different work experiences kept me thinking about the effectiveness of development projects, their success factors and constraints. When the opportunity arose to study this topic in-depth through a PhD at Unesco-IHE, I decided to go for it.

During my four and a half years at Unesco-IHE, I have thoroughly enjoyed the PhD journey. Although difficult at times, I have learned plenty about my topic of study, and moreover, I have learned so much about myself as well. For the end result, I would like to thank a number of people.

First of all I would like to express my gratitude and appreciation to my promoter, Professor Guy Alaerts. He provided me with many ideas concerning the direction and set-up of my study, gently encouraged me every time and has always been patient with me.
Jill Slinger, my copromoter, was not only a very good coach for my PhD, but also taught me many things about life in general. I definitely enjoyed all our conversations.
I am also grateful to Jan Luijendijk, former Head of the Department of Hydro-Informatics and Knowledge Management, for giving me the chance to do this work, his hospitality in Ridderkerk, for his endless enthusiasm on my topic of study and support to my work.
Jetze Heun, it was wonderful to receive a 14-page handwritten letter about my PhD proposal from your holiday address! I sincerely thank you for your constructive criticism and concern for my well-being.
Carel Keuls, I enjoyed the time we shared an office. Thank you for sharing your taste in music and for solving nearly all my IT problems.
Corrie de Haan, thank you for your help in keeping overview and in helping me to finish this project.

I would furthermore like to express my gratitude to Maarten Hofstra and Jan Leentvaar for their valuable comments on the chapter about the Dutch water sector. I want to thank Rozemarijn ter Horst for her help during her internship at Unesco-IHE, and Eelco Harteveld and Mark Hessels for their assistance with my statistical analyses.

I am indebted to many informants who shared their costly time with me to clarify the functioning of the Indonesian and Dutch public water sector. Particularly I am grateful for the support I received at the Ministry of Public Works, Directorate General of Water Resources in Indonesia.

Likewise I want to mention my fellow PhD friends, Maria Pascual, Silas Mvulirwenande, Gerald Corzo Perez, Assiyeh Tabatabai, Jeltsje Kemerink and Hermen Smit: thank you for your friendship and enlightening discussions about PhD's and life in general. The 'submission' party is well worth remembering!

Then, I would like to thank my family, especially my parents, and friends inside and outside Unesco-IHE for their support (and distraction!) along the way and for your patience when you didn't see me around for long periods of time, especially in the last year.
Gert-Jan and Nynke, thank you for your great writing facilities in Lettelbert and Wageningen.
Hermen, I have so enjoyed our writing retreats, they were conducive for my PhD and for our joint capacity development in running.

Maarten en Hermen, I am happy and grateful that you will stand by me during my defense.

Lastly, I want to thank Karel. Thank you for your support during this long commitment, and for reminding me every once in a while that there's more to life than my PhD. With Sylke, we have now started a new adventure.

The Hague, March 2013   Judith Kaspersma

13 Curriculum Vitae

Family Name:	Kaspersma
First names:	Judith Machteld
Degree:	Master of Science
Permanent	Buijs Ballotstraat 96,
Address:	2563 ZN, The Hague
	The Netherlands
Telephone:	
Number:	+31615628814
E-mail address:	j.kaspersma@unesco-ihe.org, Judith_Kaspersma@hotmail.com
Date of birth:	September 2, 1976
Place of birth:	Winschoten, The Netherlands
Nationality:	Dutch
Sex:	Female
Children:	1 daughter

Key qualifications

- Development and application of capacity development theory,
- Development and implementation of capacity needs assessments,
- Project coördination and implementation
- Consultancy, training of trainers
- Qualitative and quantitative social science research methods

Education

2008 – present	UNESCO-IHE, Delft/TU Delft
	Degree: finalising a PhD degree (by 15 September 2012) in knowledge and capacity development for the water sector
1995 - 2001	Wageningen University and Research Centre in Wageningen
	Degree: Master of Science in water resources management and irrigation engineering, http://www.iwe.wur.nl/UK/
1988 - 1994	Dutch Highschool system VWO (advanced scientific education), College Noetsele in Nijverdal, the Netherlands.

Relevant post-academic training

- Training Proposal writing for the EU ERC grant, May 2012
- Academic writing skills, course at Delft University of Technology, September - November 2010
- 3 day PhD seminar, February 2010, Delft University of Technology, Graduate school of the Faculty of Technology and Policy Analysis
- Institutional Development and Organisational Strengthening, MDF

Sri Lanka, 22 Aug – 2 Sep 2005
- Project management skills for professionals in international organisations, MDF Sri Lanka, 7 – 11 March 2005
- Logical Framework and Proposal writing for international projects. MDF Netherlands, 5 – 9 January 2004.

Country experience

Bangladesh, East-Timor, Egypt, India, Indonesia, Iran, Jordan, Pakistan, South-Africa, USA, Yemen.

Professional experience

March 2008 – present	**Organisation:** UNESCO-IHE, Delft, The Netherlands **Position:** researcher **Responsibilities:** development of capacity development theory, PhD research on the effect of a capacity development intervention in the public water sector in Indonesia, in comparison with a case in the Netherlands, where many capacity development mechanisms are presumably available. Developing capacity needs assessments, labour market studies in a range of capacity development projects in the water sector.
November 2006 – March 2008	**Organisation:** ICCO, Interchurch Organisation for Development Cooperation. **Position:** programme officer **Resposibilities:** Project-monitoring and evaluation and contribution to capacity development of organisations working on basic health care in Indonesia and East-Timor.
March – October 2006	**Organisation:** Netherlands Water Partnership, Delft, The Netherlands **Position:** Project officer, Strategic Marketing, Information & Communication **Responsibilities:** Organisation of network meetings, facilitate establishment of business consortia in the water sector.
December 2002 – February 2006	**Organisation:** Arcadis Euroconsult BV., Arnhem, The Netherlands **Position:** One continuous assignment as a junior professional officer (JPO) in consecutively the South Java Flood Control Sector Project, Indonesia, Northern Sumatra Irrigated Agriculture Project, Indonesia, Maintenance of Drains Phase II project, Egypt, and Irrigation

Improvement Project , Yemen. Part-time interim country representative for regional office in Jakarta, Indonesia (2004).

Responsibilities: Study of the Water use rights at river basin level in Northern Sumatra, assist in the establishment of 2 river basin management organisations in Northern Sumatra, design and implementation of a Management Information System for the Drainage Authority of Egypt, a Training of Trainers on Management Information Systems in Egypt.

February -
October 2002

Employer: Wageningen UR, Irrigation and Water Engineering group
Country: The Netherlands
Description of work: proposal writing, organisation of workshops
Position: research assistant
Responsibilities: Formulation of an EU-project proposal on the management of scarce regional water resources in the Mediterranean area. Organisation of a workshop on 'the Use of Appropriately treated domestic wastewater in irrigated agriculture – Technical and Socio-Economic Aspects'. Research on the development of guidelines for appropriate agricultural use of treated effluent.

Experience during MSc degree

August 2000 -
April 2001

Employer: Wageningen UR/Water and Environment Research and Study Centre in Amman, Jordan.
Country: Jordan
Position: Research assistant, MSc thesis research
Responsibilities: Modelling drip irrigation with treated wastewater in Jordan, resulting in recommendations for improved water management and agricultural production in the Jordan valley.

September 1999 -
January 2000.

Employer: Texas Agricultural Experiment Station in Lubbock.
Country: USA
Position: Research assistant, MSc thesis research
Responsibilities: Establish criteria for the definition of management units for precision irrigation, with Arcview GIS and SAS. Thesis received grade 9 out of 10.

March – September 1998	**Employer:** Arcadis Euroconsult BV. **Country:** Pakistan **Client:** Soil Survey of Pakistan **Project title and description:** 'Strengthening the Soil Survey of Pakistan.' **Position:** intern **Responsibilities:** Salinity field research, training Soil Survey officers, writing a manual for the use of field instruments.

Relevant skills

Software Proficiency:	Atlas.TI: software for qualitative data analysis Microsoft Office applications Hymos 4 Hydrological model/database
Languages:	**Dutch:** mother tongue **English:** fluent **French:** moderate **German:** moderate **Indonesian:** good **Arabic::** basic

Recent and/or relevant congresses, symposia, excursions

May 2012	IWA Water, Climate and Energy, presented a paper
April 2012	Developmental Evaluation, seminar at Wageningen UR
March 2009	5th World Water Forum, Istanbul
July 9, 2008 &	Application of Complexity Theory in International Aid, ODI London, October 3, 2008, UK.
October, December 2004	Workshops for the establishment of river basin management organisations for Jambi and Kampar river basins in Northern Sumatra, Indonesia.

Hobbies

Singing
Running
Hiking, camping
Photography

Kaspersma, J. M., Alaerts, G. J., and Slinger, J. H.: Competence formation and post-graduate education in the public water sector in Indonesia, Hydrol. Earth Syst. Sci., 16, 2379-2392, 2012

Kaspersma, J. M., Alaerts, G. J., and Slinger, J. H.: Readiness for future challenges: Organic vs. mechanic organisational structure at the DGWR in Indonesia, World Congress on Water, Climate and Energy - Building a sustainable global future, Dublin, Ireland, 2012, 13.

Alaerts, G. J., and Kaspersma, J. M.: Progress and challenges in knowledge and capacity development, in: Capacity Development for improved water management, edited by: Blokland, M. W., Alaerts, G. J., Kaspersma, J. M., and Hare, M., Taylor and Francis, Delft, 327, 2009.

Blokland, M. W., Alaerts, G. J., Kaspersma, J. M., and Hare, M. Eds. Capacity Development for improved water management. Taylor and Francis, Delft, 327, 2009.

T - #0407 - 101024 - C38 - 240/165/12 - PB - 9781138000971 - Gloss Lamination